Springer Geology

The book series Springer Geology comprises a broad portfolio of scientific books, aiming at researchers, students, and everyone interested in geology. The series includes peer-reviewed monographs, edited volumes, textbooks, and conference proceedings. It covers the entire research area of geology including, but not limited to, economic geology, mineral resources, historical geology, quantitative geology, structural geology, geomorphology, paleontology, and sedimentology.

More information about this series at http://www.springer.com/series/10172

Ivar Murdmaa · Elena Ivanova
Editors

The Ioffe Drift

Springer

Editors
Ivar Murdmaa
Shirshov Institute of Oceanology
Russian Academy of Sciences
Moscow, Russia

Elena Ivanova
Shirshov Institute of Oceanology
Russian Academy of Sciences
Moscow, Russia

ISSN 2197-9545 ISSN 2197-9553 (electronic)
Springer Geology
ISBN 978-3-030-82870-7 ISBN 978-3-030-82871-4 (eBook)
https://doi.org/10.1007/978-3-030-82871-4

This Springer imprint is published by the registered company Springer Nature Switzerland AG
The registered company address is: Gewerbestrasse 11, 6330 Cham, Switzerland

Preface

The discovery of a calcareous contourite drift overlying the Florianopolis Fracture Zone ridge in the western South Atlantic during cruise 32 (2010) of the research vessel (R/V) *Akademic Ioffe* opened up new approaches and perspectives in the contourite theory. Although demonstrating similar behavior relative to bottom-water dynamics, rather rare and understudied calcareous contourites differ from their terrigenous analogs in terms of the origin, grain-size distribution, chemical and mineral composition of sediment particles. Therefore, a detailed multidisciplinary study of the Ioffe Drift provides new knowledge on biogenic contourites deposited in the pelagic realm, in conditions of low biological productivity and terrigenous material supply, under the influence of the Antarctic Bottom Water (AABW) and Lower Circumpolar Deep Water (LCDW) flow from the Vema Channel. As shown by our study, the contour currents associated with LCDW and AABW create numerous erosional hiatuses in the Ioffe Drift, thus increasing its uniqueness and difference from the better studied North Atlantic drifts.

The use of a return track of the R/V *Akademic Ioffe* during cruises 32, 46 and 52 for short-term scientific investigations permitted a few lines and a detailed mini-polygon of high-resolution seismoacoustic survey across the Ioffe Drift and in its summit area, respectively, to a sub-bottom depth of 30–60 m. Six sediment cores retrieved from the drift area recovered the Upper Pliocene–Quaternary deposits. Two grab samples with ferromanganese nodules were obtained from the channel in the transform fracture zone to the north of the drift. Along with the materials collected from the western South and Central Atlantic during seven other cruises of the same vessel, organized by the Shirshov Institute, Russian Academy of Sciences, the results of the multi-proxy study of the Ioffe Drift served as the basis for Project 18-17-00227 "The Pliocene–Quaternary contourites of the Central and Southern Atlantic", funded by the Russian Science Foundation (RSF) from 2018 to 2020. This monograph presents the summarized original scientific results obtained to date on the Ioffe Drift, mostly from the aforementioned RSF project, and partially within the State assignments № 0149-2019-0007 and 0128-2021-0006.

The authors are grateful to the scientific parties and the masters and crews of R/V *Akademik Ioffe* on cruises 32, 46 and 52 for their professional support in the collection of scientific materials: L. Reykhard, I. Mullabaev, N. Nemchenko, N. Simagin, I. Klementieva, A. Adamovich, L. Demina and M. Kornilova are thanked for technical assistance with the ferromanganese nodules study and measurements of magnetic susceptibility, color reflectance, grain-size and $CaCO_3$. M. Zenina and A. Boev are acknowledged for professional support with the SEM images. Technical support with figures and SEM plates was given by S. Sim, E. Streltsova, K. Ginzburg, M. Vasil'eva and N. Grechikhina. We appreciate the inspiring scientific discussions held with Gregoriy Barenblatt, Nick McCave, Galina Kazarina, Bruno Galbrun, Eva Moreno and Ekaterina Ovsepyan.

Our special thanks go to the Russian Science Foundation and Shirshov Institute of Oceanology for funding the monograph's preparation and for the ship time, respectively.

Moscow, Russia
March 2021

Ivar Murdmaa
Elena Ivanova

Contents

Introduction

Elena Ivanova, Ivar Murdmaa, and Dmitrii Borisov

Abstract The Ioffe Drift has a special place among the discoveries of the contourite and mixed contourite-turbidite depositional systems of the western South Atlantic along the path of the Antarctic Bottom Water (AABW) and Lower Circumpolar Deep Water (LCDW) into the Brazil Basin. The drift is located in the Southern Brazil Basin, to the north of Rio Grande Rise and north-east of the northern exit from the Vema Channel, in the core of a large oligotrophic gyre with an anticlockwise circulation of largescale, upper-level geostrophic currents and low biological productivity. The terrigenous material supply to the drift area is rather low due to its remoteness from the continent and dominant directions of the south-east trade winds. The sediments in the drift area are mostly composed of calcareous microfossils. A multiproxy study of the sediment cores along with the seismoacoustic records provides multiple evidence of the contourite origin of the drift.

The Ioffe Drift has a special place among the discoveries of the contourite and mixed contourite-turbidite depositional systems of the western South Atlantic along the path of the Antarctic Bottom Water (AABW) and Lower Circumpolar Deep Water (LCDW) into the Brazil Basin. These contourites include the Vema contourite fan and the mixed systems of the Columbia Channel and São Tomé Seamount (e.g., Mézerais et al. 1993; Massé et al. 1994; Lima et al. 2009; Borisov et al. 2013).

The Ioffe Drift is located in the Southern Brazil Basin, to the north of Rio Grande Rise and north-east of the northern exit from the Vema Channel, in the core of a large oligotrophic gyre with an anticlockwise circulation of large-scale, upper-level geostrophic currents (Fig. 1; see also Chap. 2). The relatively warm subtropical waters of the gyre are characterized by low biological productivity of the pelagic

E. Ivanova (✉) · I. Murdmaa · D. Borisov
Shirshov Institute of Oceanology of the Russian Academy of Sciences, Moscow, Russia
e-mail: e_v_ivanova@ocean.ru

I. Murdmaa
e-mail: murdmaa@mail.ru

D. Borisov
e-mail: dborisov@ocean.ru

I. Murdmaa and E. Ivanova (eds.), *The Ioffe Drift*, Springer Geology,
https://doi.org/10.1007/978-3-030-82871-4_1

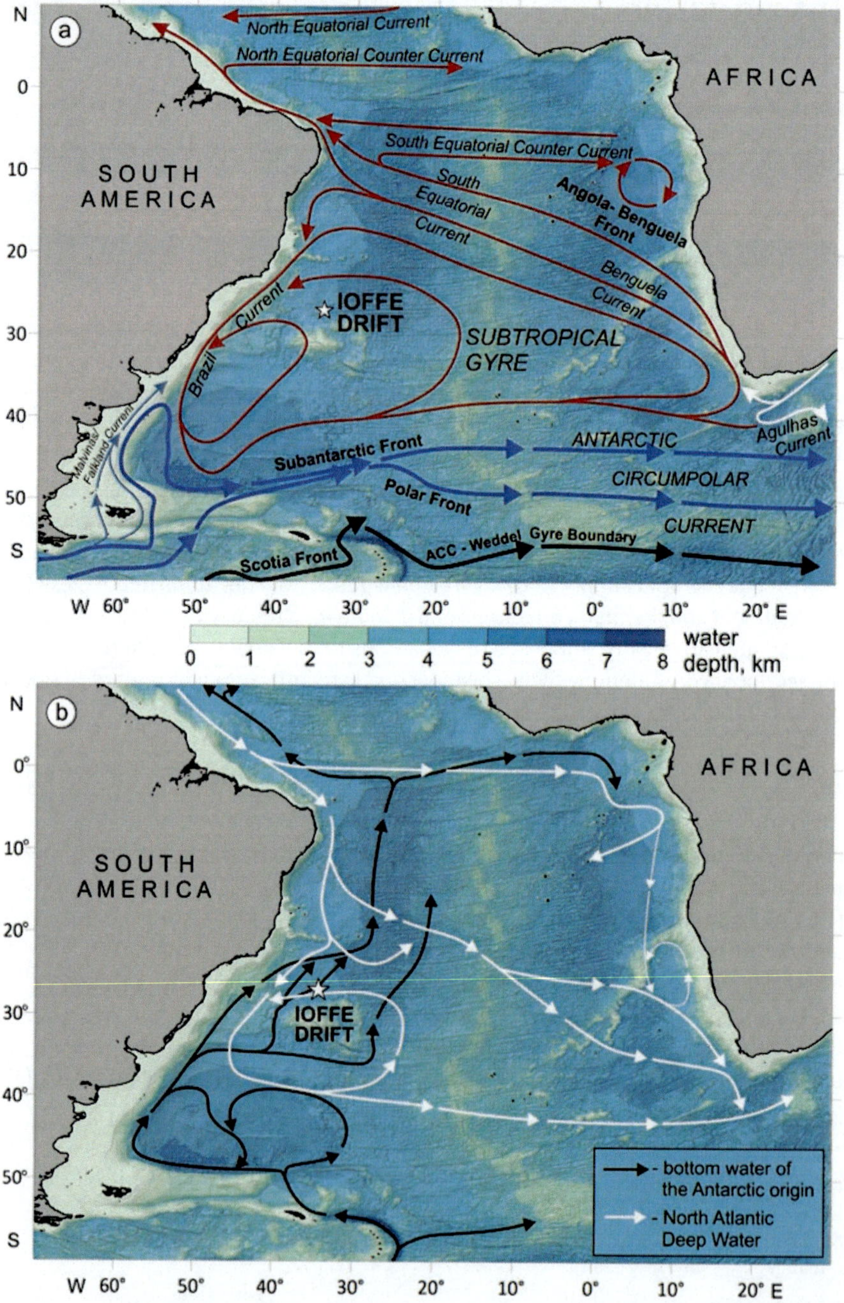

Fig. 1 General scheme of large-scale geostrophic currents and fronts in the Central and South Atlantic: **a** Upper-level circulation (Berger and Wefer 1996); **b** bottom-level circulation in the same region (Morozov et al. 2010)

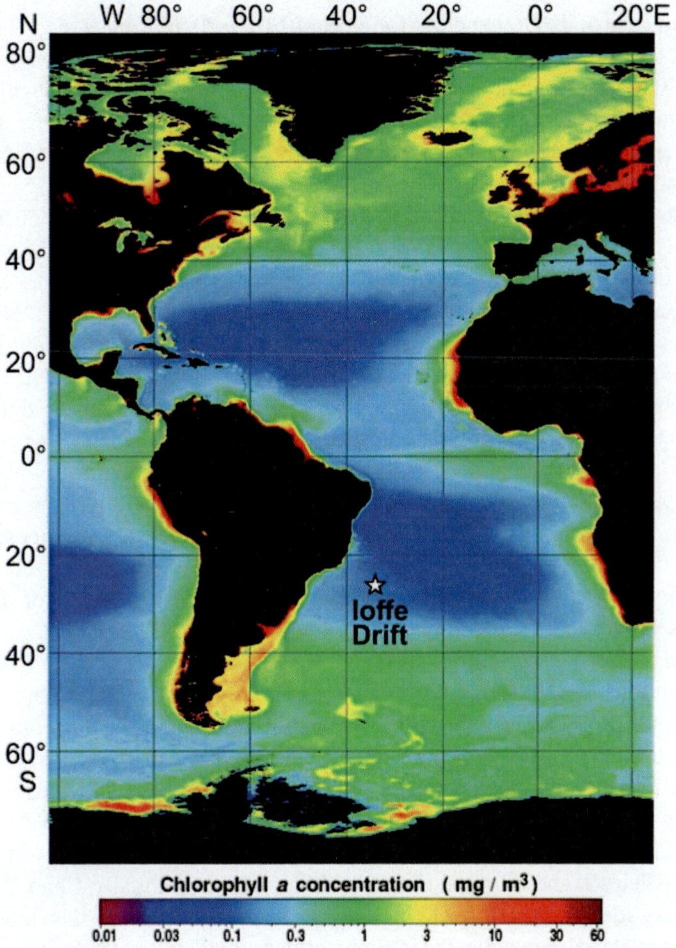

Fig. 2 Mean annual sea surface chlorophyll-*a* concentration in the Atlantic Ocean for the period 2012–18 (satellite data from the Moderate Resolution Imaging Spectroradiometer (MODIS)) (NASA Ocean Biology 2014)

type (as suggested by, for instance, the chlorophyll-*a* concentration in Fig. 2). The terrigenous material supply to the drift area is rather low due to its remoteness from the continent and dominant directions of the south-east trade winds. Therefore, recent deep-sea sediments in the drift area above the calcite (and foraminiferal) lysocline are mostly composed of calcareous microfossils, namely planktic foraminifers and nannofossils and their fragments.

Large-scale deep- and bottom-water circulation in the area is controlled by the flows of LCDW and AABW, respectively (Morozov et al. 2010). Contour currents of these water masses of Antarctic origin (Fig. 1b; Morozov et al. 2010) are known to play an important role in the accumulation and erosion of contourites in the western

South Atlantic (e.g., Hernández-Molina et al. 2010, 2016; Preu et al. 2013; Ivanova et al. 2016, 2020). A major part of the drift body is located at a water depth of 3,800–4,000 m; that is, shallower than the LCDW/AABW boundary, roughly defining the foraminiferal lysocline (Ruddiman and Heezen 1967; Berger 1968, Orsi et al. 1999; Hernández-Molina et al. 2016; see Chap. 2). Therefore, the preservation of its calcareous components generally ranges from good to perfect.

Previous studies have demonstrated a wide distribution of hiatuses disturbing the continuity of the geological record in the Ioffe Drift, and thus hindering the stratigraphic investigation based on the existing stratigraphic frame and the correlation of contourite sections in the western South Atlantic (Ivanova et al. 2016, 2020; see Chaps. 3, 6, 7 and 9). Meanwhile, in terms of geological time, the identification of temporal gaps in the contourite records are of special importance to the interpretation of changes in sedimentation, microfossil assemblages, climate and deep-water paleocirculation at different timescales (e.g., Meyers and Sageman 2004; Viana and Rebesco 2007). In this context, the calcareous Ioffe Drift offers an unique opportunity to apply a multiproxy approach, including classical biostratigraphy and modern magnetic susceptibility (MS), color reflectance and XRF scanning techniques, to identify the age of the stratigraphic units and hiatuses in its body (see Chaps. 6, 7 and 9). The grain-size distribution provides convincing evidence of the bottom current's influence on the sedimentation in the drift (see Chap. 6). Occurrence of ferromanganese nodules and micronodules supports the pelagic realm with a low-sedimentation rates and oxic environment of the sedimentation (see Chap. 8). Collectively, a multiproxy study of the sediment cores along with the seismoacoustic records provides multiple evidence of the contourite origin of the drift.

Thus, this monograph sheds new light on the nature of a previously unknown calcareous drift, including its seismic structure and bathymetry (Chap. 5), sedimentology (Chap. 6), biostratigraphy (Chap. 7), occurrence and features of erosional hiatuses (Chap. 9) and ferromanganese nodules (Chap. 8). Finally, Chap. 10 presents preliminary ideas on the Ioffe Drift history. The latter has to be developed further when new materials are collected in future cruises. In addition, Chap. 2 describes the physiographic and geological setting, Chap. 3 demonstrates the development of the regional chronostratigraphic framework, while Chap. 4 considers the materials and general methods used in the study.

References

Berger WH (1968) Planktonic foraminifera: selective solution and paleoclimatic interpretation. Deep-Sea Res 15:31–43

Berger WH, Wefer G (1996) Expeditions into the past: paleoceanographic studies in the South Atlantic. In: Wefer G, Berger WH, Siedler G, Webb DJ (eds) The South Atlantic: present and past circulation. Springer

Borisov DG, Murdmaa IO, Ivanova EV et al (2013) Contourite systems in the region of the Southern San Paulo Plateau escarpment, South Atlantic. Oceanology 53:460–471 (in Russian with English trans). https://doi.org/10.1134/S1028334X13090146

Hernández-Molina FJ, Paterlini M, Somoza L et al (2010) Giant mounded drifts in the Argentine continental margin: origins, and global implications for the history of thermohaline circulation. Mar Pet Geol 27:1508–1530. https://doi.org/10.1016/j.marpetgeo.2010.04.003

Hernández-Molina FJ, Soto M, Piola AR et al (2016) A contourite depositional system along the Uruguayan continental margin: sedimentary, oceanographic and paleoceanographic implications. Mar Geol 378:333–349. https://doi.org/10.1016/j.margeo.2015.10.008

Ivanova E, Borisov D, Dmitrenko O, Murdmaa I (2020) Hiatuses in the late Pliocene-Pleistocene stratigraphy of the Ioffe calcareous contourite drift, western South Atlantic. Mar Pet Geol 111:624–637. https://doi.org/10.1016/j.marpetgeo.2019.08.031

Ivanova E, Murdmaa I, Borisov D, Dmitrenko O, Levchenko O, Emelyanov E (2016) Late Pliocene-Pleistocene stratigraphy and history of formation of the Ioffe calcareous contourite drift. Western South Atlantic Mar Geol 372:17–30. https://doi.org/10.1016/j.margeo.2015.12.002

Lima AF, Faugères JC, Mahiques M (2009) The Oligocene–Neogene deep-sea Columbia channel system in the South Brazilian Basin: seismic stratigraphy and environmental changes. Mar Geol 266:18–41. https://doi.ogr/https://doi.org/10.1016/j.margeo.2009.07.009

Massé L, Faugères JC, Bernat M et al (1994) A 600,000-year record of Antarctic bottom water activity inferred from sediment textures and structures in a sediment core from the Southern Brazil Basin. Paleoceanography 9:1017–1026. https://doi.org/10.1029/94PA01442

Meyers SR, Sageman BB (2004) Detection, quantification, and significance of hiatuses in pelagic and hemipelagic strata. Earth Planet Sci Lett 224:55–72

Mézerais ML, Faugères JC, Figueiredo AG, Massé L (1993) Contour current accumulation off the Vema Channel mouth, southern Brazil Basin: pattern of a "contourite fan." Sediment Geol 82:173–187. https://doi.org/10.1016/0037-0738(93)90120-T

Morozov EG, Demidov AN, Tarakanov RY, Zenk W (2010) Abyssal channels in the Atlantic Ocean. Springer, Netherlands, Dordrecht

NASA Ocean Biology (OB.DAAC) (2014) Mean annual sea surface chlorophyll-*a* concentration for the period 2012–2018. Data obtained from the moderate resolution imaging spectroradiometer (MODIS) aqua ocean color website. NASA OB.DAAC, Greenbelt, MD. http://oceancolor.gsfc.nasa.gov/cgi/l3

Orsi AH, Johnson GC, Bullister JL (1999) Circulation, mixing, and production of Antarctic bottom water. Prog Oceanogr 43:55–109. https://doi.org/10.1016/S0079-6611(99)00004-X

Preu B, Hernández-Molina FJ, Violante R et al (2013) Morphosedimentary and hydrographic features of the northern Argentine margin: the interplay between erosive, depositional and gravitational processes and its conceptual implications. Deep Res Part I Oceanogr. Res Pap 75:157–174. https://doi.org/10.1016/j.dsr.2012.12.013

Ruddiman WF, Heezen BC (1967) Differential solution of planktonic foraminifera. Deep Sea Res 14:801–808

Viana AR, Rebesco M (eds) (2007) Economic and palaeoceanographic importance of contourite deposits. Geol Soc (London, Special publication) 276:119–224

Regional Setting

Dmitrii Borisov, Ivar Murdmaa, Oleg Levchenko, and Dmitry Frey

Abstract The focus area of this monograph is the central part of a tectonic ridge within the Florianopolis (Rio Grande) Fracture Zone in the southern Brazil Basin. The ridge is swept by the bottom current of the Lower Circumpolar Deep Water propagating from the Argentine Basin to the Brazil Basin through the Vema Channel.

1 Physiography

The Brazilian Margin between 35° and 20°S (apart from the São Paulo Plateau) is a typical passive margin (Mascle 1976; Chang et al. 1992) resulting from the breakup of Gondwana during the Mesozoic and the subsequent separation of the South American and African continents. The margin is characterized by a wide continental shelf (up to 220 km) demonstrating a slope angle from 0°27′ to 0°59′ and a maximum depth of 300 m (Campos et al. 1974; Bassetto et al. 2000). The relatively narrow and steep continental slope (3–5°) extends at a range of depths: 100–300 m at the shelf break and 1,000–3,000 m at its outer boundary (Bassetto et al. 2000). The continental slope is locally interrupted by plateaus, terraces, erosional channels, slide deposits, deep-sea fans and sedimentary drifts (Jeck et al. 2019). This area represents the transition between the Argentine Basin to the south and the Brazil Basin to the north. Its most prominent physiographic features are the Santa Catarina Plateau, the São Paulo Plateau (with the São Paulo Ridge), the Rio Grande Rise, the Rio Grande Terrace, the Pelotas Drift, the Ewing Drift, the Vema Channel, the Columbia Channel and the Vitoria–Trindade seamounts (Fig. 1a). Other important regional features located to the east of the São Paulo Plateau are the Jean Charcot and Kazanskiy seamounts. The

D. Borisov (✉) · I. Murdmaa · O. Levchenko · D. Frey
Shirshov Institute of Oceanology, Russian Academy of Sciences, Moscow, Russia
e-mail: dborisov@ocean.ru

I. Murdmaa
e-mail: murdmaa@mail.ru

O. Levchenko
e-mail: olevses@mail.ru

I. Murdmaa and E. Ivanova (eds.), *The Ioffe Drift*, Springer Geology,
https://doi.org/10.1007/978-3-030-82871-4_2

seafloor in the area has been affected by several remarkable crustal discontinuities demonstrating a general E–W trend, corresponding to fracture zones and oceanic lineaments. The most prominent are the Rio de Janeiro Fracture Zone, the Rio Grande Fracture Zone (also known as the Florianopolis Fracture Zone) and the Porto Alegre and Chui Lineaments (Fig. 1a). The Cruzeiro do Sul (Southern Cross) Fracture Zone and Capricórnio Lineament are oriented in a NW–SE direction and run across the São Paulo Plateau, the Santa Catarina Plateau and the Rio Grande Rise (Bassetto et al. 2000).

The Santa Catarina Plateau is traced at a depth of 2,900–4,000 m. It extends for 450 km from north to south and 700 km from east to west (Jeck et al. 2019). The Rio Grande Terrace interrupts the upper continental slope at a depth of 650–2,900 m. Its southern and eastern boundaries are marked by erosional channels formed under the influence of bottom currents. The Torres High is one of the basement highs overlaid by the Santa Catarina Plateau (Kowsmann and Leyden 1977; Fontana 1996). It is oriented in a general NW–SE direction between latitudes 29° and 31°S at depths ranging from 3,400 to 3,900 m (Jeck et al. 2019).

The extensive São Paulo Plateau is located on the continental slope at a depth of 2,000–3,200 m (Viana et al. 2003) (or 1,200–3,200 m, according to Jeck et al. 2019). The plateau is affected by numerous halokinetic structures related to the Aptian salt deposits. Active diapiric structures control the courses of the many channels that form a complex network over the plateau (Viana et al. 2003). In the south-western part, there is a large incision (submarine valley) oriented NW–SE, collecting the surrounding submarine drainage network (Duarte and Viana 2007). This valley is connected upslope to the Cananéia canyon, which breaches the shelf, apparent at the 50 m contour (Duarte and Viana 2007). According to Bueno et al. (2004), this submarine drainage coincides with the Capricórnio Lineament. The plateau is bounded to the south by a deep linear valley related to the Florianopolis Fracture Zone. The height of the escarpment varies from 500 m near the Rio Grande Terrace to more than 1,500m in the area of the São Paulo Ridge, extending along the southern boundary of the plateau.

The Rio Grande Rise is located between 28° and 34°S and between 28° and 40°W. It extends for approximately 400 km from east to west and 650 km from north to south (along the 3,500 m contour) (Barker et al. 1983b; Gamboa et al. 1983). The rise has a height of over 4 km relative to the adjacent seafloor and is divided into two segments by a clearly expressed depression. The eastern segment has a U-shaped form, while the western is nearly oval (Camboa and Rabinowitz 1984). The Cruzeiro do Sul Trough intersects the rise that follows the eponymous lineament. The trough comprises a graben oriented NW–SE with a relative depth of 1,400–1,600 m, formed by tectonic activity in the Middle Eocene (Barker et al. 1983b). On the smooth top of the rise, the trough is framed by two guyot chains characterized by flat tops at depths of 560–900 m. The steeper northern slope of the Rio Grande Rise is interrupted by canyons, while the southern and western slopes are gentler.

The Vema Channel occupies the deepest part of the Rio Grande Gap (at a water depth of over 4,800 m), separating the Rio Grande Rise from the Santa Catarina and São Paulo Plateaus. The channel connects the Argentine and Brazil Basins and plays a crucial role in the northward distribution of Antarctic Bottom Water. The channel follows the general S–N trend and extends for 700 km. Its orientation is mostly controlled by basement relief (Johnson 1984). The channel width is approximately 20–30 km at the prominent slope-break near 4,300 m. It becomes wider at the northern exit and, north of 29°30′S, turns slightly toward the northeast.

The Hunter Channel is another major gateway for the distribution of the Antarctic Bottom Water to the north. It is a 200 km wide bathymetric depression located east of the Rio Grande Rise at depth of 4,300 m (Pätzold et al. 1996).

The seafloor to the east of the São Paulo Plateau is dissected by several channels oriented normal to the regional contours (Gonthier et al. 2003). The most prominent is the Columbia (Trindade) Channel, which has a depth of 200–400 m and a width of up to 20 km. This runs in a WNW–ESE direction at a depth range of 4,200–5,000 m (Faugères et al. 2002b; Lima et al. 2009). The Rio de Janeiro Channel is much smaller. It is oriented in a W–E direction and is linked to the Rio de Janeiro Fracture Zone (Bassetto et al. 2000; Gonthier et al. 2003). The largest tributaries of the Columbia and Rio de Janeiro Channels are the Maçae Channel, running along the foot of the Almirante Saldanha seamount, and the Carioca Channel, respectively.

The Vitoria–Trindade seamount chain is a narrow, linear structure extending along 20°30′S from the continental slope toward the east for 1,200 km. This structure is divided into western, central and eastern segments. The western segment consists of a guyot chain. The central segment includes the Dogaressa Bank, Jaseur and Davis seamounts. The eastern segment of the chain embraces several separate seamounts, including the Trindade and Martin Vaz Islands (Skolotnev et al. 2010). The origin of this chain is often attributed to the activity of the Trindade hotspot (Duncan and Richards 1991; Fodor and Hanan 2000).

The study area is located in the central part of the narrow ridge within the Rio Grande (Florianopolis) Fracture Zone. The ridge is oriented in a SW–NE direction (strike angle 70–73°), at a sharp angle to a series of valleys and associated ridges of tectonic origin. The ridge is traced between 26°19′S and 27°40′S and between 36°08′W and 31°30′W. It has steep northern slopes and gentler southern slopes. The length of the ridge crest is approximately 470–500 km. Its widest point is 90 km at 33°55′W. To the north, the ridge is bounded by a deep valley (300 m relative depth in a water depth of 4,800 m) connected to the Vema Channel. This valley is interrupted by a ledge protruding from the ridge slope to the north at 33°54′W. The ridge rises more than 700 m above the adjacent seafloor and its summit is at a water depth of 3,800 m.

2 Oceanography

Interaction between highly active oceanographic processes and the seafloor, with a strong impact of these processes on sedimentation, is an essential characteristic of the Atlantic South American Margin (Hernández-Molina et al. 2010, 2016; Preu et al. 2013; Frey et al. 2018). The region's oceanographic regime is affected by the interaction of several major water masses: Tropical Water (TW); South Atlantic Central Water (SACW); Antarctic Intermediate Water (AAIW); Upper Circumpolar Deep Water (UCDW); North Atlantic Deep Water (NADW); Lower Circumpolar Deep Water (LCDW) and Antarctic Bottom Water (AABW).

This study uses the neutral density as a parameter for distinguishing (Jackett and McDougall 1997) the boundaries between the water masses, following Orsi et al. (1999), Jullion et al. (2010) and Hernández-Molina et al. (2016). The Brazil Current generally controls surface circulation over the Brazilian Southern Margin. It is a western boundary current associated with the anticyclonic South Atlantic Subtropical Gyre , transporting warm and saline TW and SACW at depth interval from the ocean surface to 430 m toward the southwest (Piola and Matano 2001) (Fig. 1a). The lower boundary of SACW corresponds to the isoline of neutral density of $\gamma^n = 26.7$ g/cm^3 (Hernández-Molina et al. 2016). SACW overlies the low-salinity AAIW that propagates to the north in a system of gyres. The north-directed AAIW flow turns to the east near 35°S and follows the southern hemisphere anticyclonic subtropical gyre. After crossing the Mid-Atlantic Ridge and moving along the western African Margin, this flow turns to the west and reaches the South American margin near 22°S. At this point, it splits into two branches (Larqué et al. 1997; Schmid et al. 2000). One branch flows to the north and the other (the so-called recirculated AAIW, or rAAIW) moves to the south, staying in contact with the continental slope up to 29°S (Hernández-Molina et al. 2016).

The boundary between the AAIW and the underlying UCDW is located at a depth of 950–1,150 m and corresponds to isolines of neutral density of $\gamma^n = 27.5$–27.55 g/cm^3 (Jullion et al. 2010; Hernández-Molina et al. 2016). The south-flowing, highly saline, oxygen-rich and nutrient-poor NADW splits the cold Circumpolar Deep Water (CDW), which is moving in the opposite direction, separating the upper part (UCDW) from the lower (LCDW). The UCDW and LCDW are also referred to by some authors as the Upper and Lower Circumpolar Waters: UCPW and LCPW (Reid et al. 1977; Morozov et al. 2010).

In the study area, the boundaries between the UCDW, NADW and LCDW at depths of 1,700–2,000 m and 3,250–3300 m, respectively (Fig. 2) correlate to the isolines of neutral density $\gamma^n = 27.87$–27.92 g/cm^3 and 28.07–28.11 g/cm^3 (Jullion et al. 2010; Hernández-Molina et al. 2016). The abyssal circulation in the adjacent area is dominated by Antarctic Bottom Water, composed of Weddell Sea Deep Water, whose upper boundary is marked by a neutral density of $\gamma^n = 28.26$–28.27 g/cm^3 (4,250–4,500 m) (Orsi et al. 1999; Morozov et al. 2010).

LCDW generally follows the AABW flow and can penetrate to the Santa Catarina Plateau from the south, moving along the continental slope. Part of the LCDW, flowing together with AABW through the Vema Channel, turns to the west at its northern exit and moves along the southern escarpment of the São Paulo Plateau to the Santa Catarina Plateau (Morozov et al. 2010).

The AABW propagates from the Argentine Basin to the Brazil Basin mainly through the Vema and Hunter channels, located west and east of the Rio Grande Rise, respectively (Fig. 1a). The coldest part of Antarctic waters ($\theta < 0$ °C) flowing through the Vema Channel is characterized by two cores. The lower is related to the

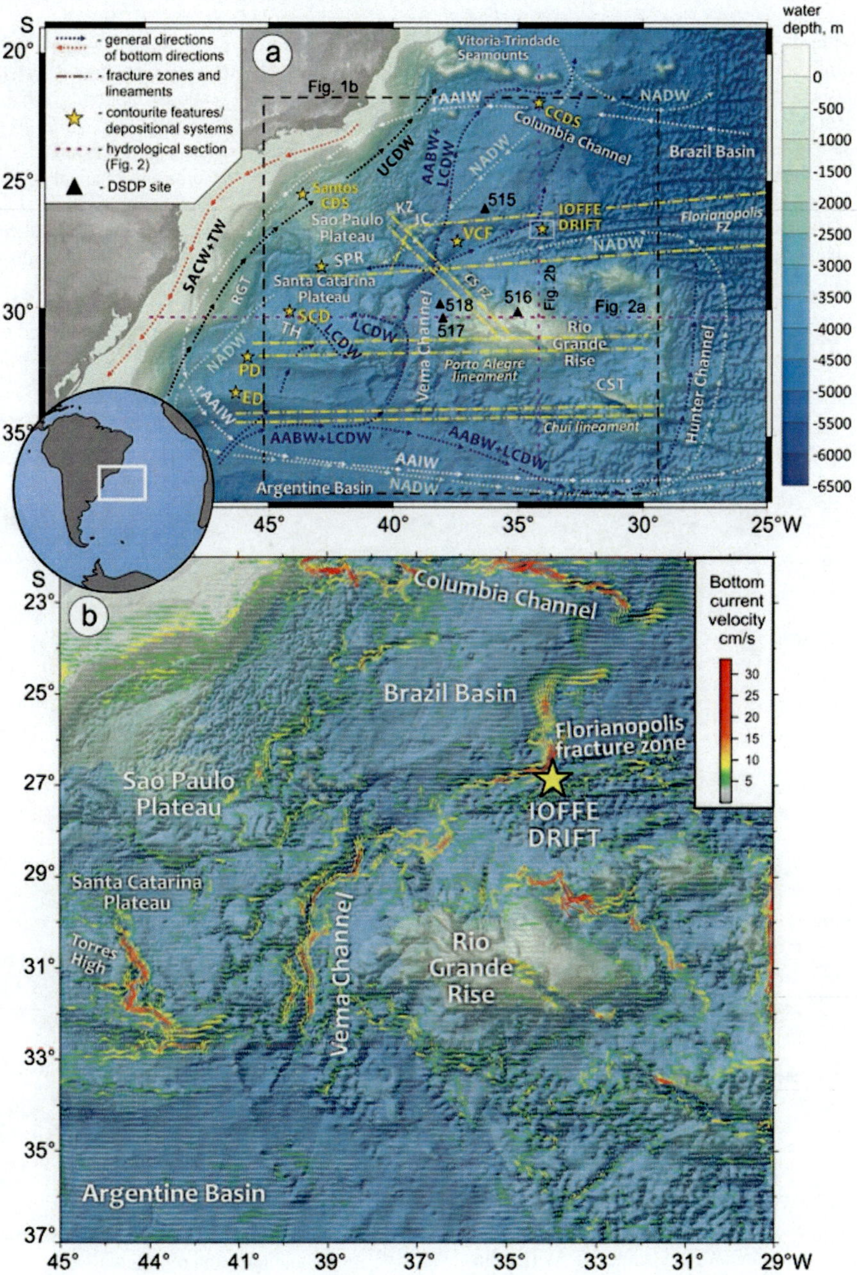
a
- general directions of bottom directions
- fracture zones and lineaments
- contourite features/ depositional systems
- hydrological section (Fig. 2)
- DSDP site
water depth, m
0
-500
-1000
-1500
-2000
-2500
-3000
-3500
-4000
-4500
-5000
-5500
-6000
-6500
Vitoria-Trindade Seamounts
Fig. 1b
AAIW
NADW
CCDS
Columbia Channel
Brazil Basin
UCDW
AABW+ LCDW
Santos CDS
515
IOFFE DRIFT
Florianopolis FZ
Sao Paulo Plateau
KZ
JC
VCF
NADW
SPR
SACW+TW
RGT
Santa Catarina Plateau
CS FZ
Fig. 2b
516
518
517
Fig. 2a
SCD
TH
LCDW
Vema Channel
Rio Grande Rise
NADW
PD
Porto Alegre lineament
CST
Hunter Channel
ED
AAIW
AABW+LCDW
Chui lineament
AAIW
Argentine Basin
NADW
S
20°
25°
30°
35°
45°
40°
35°
30°
25°W
b
Columbia Channel
Bottom current velocity cm/s
30
25
20
15
10
5
Brazil Basin
Florianopolis fracture zone
Sao Paulo Plateau
IOFFE DRIFT
Santa Catarina Plateau
Torres High
Vema Channel
Rio Grande Rise
Argentine Basin
S
23°
25°
27°
29°
31°
33°
35°
37°
45°
44°
41°
39°
37°
35°
33°
31°
29°W

◀**Fig. 1** **a** Bathymetric map of the SE Brazil margin and the adjacent abyssal plains, showing the location of major topographic features, fracture zones (FZ), lineaments, contourite depositional systems (CDS) and generalized directions of distribution of main water masses. Seafloor topography is based on the physiographic map of the Brazil oceanic margin (scale 1:1883927) (*Diretoria de Hidrografia e Navegação, Marinha do Brasil* 2017). Location of studied region is shown in inset (left). CST–Cruzeiro do Sul Trough, CS FZ–Cruzeiro do Sul fracture zone, CCDS–Columbia Channel depositional system, ED–Ewing Drift, KZ–Kazanskiy seamount, JC–Jean Charcot seamounts, PD–Pelotas Drift, RGT–Rio Grande Terrace, SCD–Santa Catarina Drift, SPR–São Paulo Ridge, TH–Torres High, VCF–Vema contourite fan. Abbreviations of water masses are given in the text. **b** Results of simulation of the near-bottom dynamics performed using INMOM model (Frey et al. 2019)

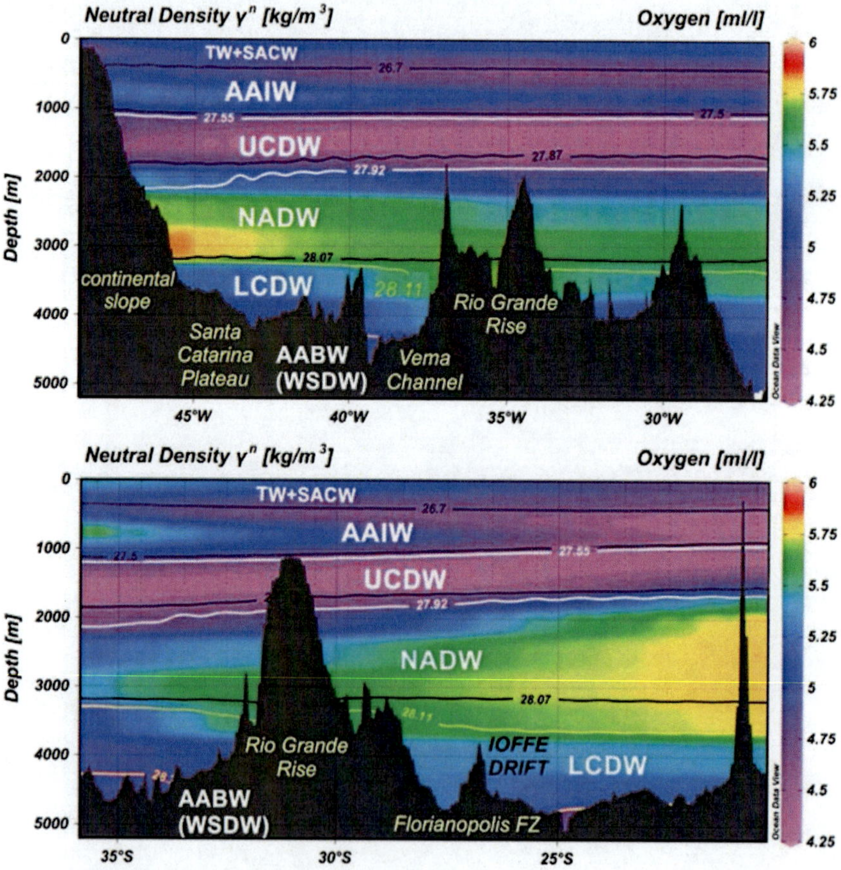

Fig. 2 Distribution of oxygen in the water and neutral density values (contour lines) over two sections along **a** 29°S; and **b** 35°W (Levitus et al. 2010), with an indication of boundaries between water masses: white lines (Jullion et al. 2010), black lines (Hernández-Molina et al. 2016), yellow lines (Orsi et al. 1999). Abbreviations of water masses are given in the text

deepest part of the channel, while the upper is at 400 m above, over its western flank. North of 34°S, the lower flow is deflected toward the northeast. In the study area it has a potential temperature of −0.087 °C and follows the northern slope of the Florianopolis Fracture Zone Ridge at a measured velocity of up to 20 cm/s. North of the study area, the coldest part of Antarctic water splits into several branches (Tarakanov and Morozov 2015). By contrast, the upper flow of Antarctic water is less intense. It is characterized by a mean flow velocity not exceeding 5 cm/s and its higher potential temperature ($\theta = -0.063$ °C) (Tarakanov and Morozov 2015). North-flowing bottom currents of Antarctic water move to the east along the Vitoria–Trindade chain and pass through the deep passages between the Dogaressa Bank, Columbia seamount and Trindade Island (Morozov et al. 2010; Frey et al. 2019). Modern AABW current velocities do not exceed 10 cm/s in the open basin, yet in channels and deep submarine valleys they exceed 30 cm/s (Frey et al. 2018). Simulation of the near-bottom dynamics performed using the Ocean Model developed by the Institute of Numerical Mathematics (INMOM, Dianskii et al. 2002) demonstrates maximum current velocities of up to 30 cm/s for the Vema Channel, Hunter Channel, the northeastern slope of the Rio Grande Rise, southern Santa Catarina Plateau (along the slopes of the Torres High), the deep valley at the base of the ridge covered by the Ioffe Drift and Columbia Channel (Fig. 1b; Frey et al. 2019).

Intense bottom currents flowing through the Vema Channel transport a large amount of suspended matter. The bottom nepheloid layer is characterized by a thickness of 500–1,000 m (Biscaye and Eittreim 1977; Morozov et al. 2010). Several intermediate nepheloid layers were also revealed on the transect from the Vema Channel to the Florianopolis Fracture Zone Ridge at water depths of 800–1,500 m, 2,200–3,500 m and 3,500–4,300 m (Morozov et al. 2010). Two nepheloid layers are recognized in the study area: an intermediate layer at a depth range of 3,400–4,200 m; and a bottom layer at a depth below 4,200 m. The volume concentration of suspended matter in these layers reaches 0.2 and 0.3 mm^3/l in the intermediate and bottom layers, respectively (Morozov et al. 2010).

The suspended matter in bottom waters flowing through the Vema Channel comprises fragments of diatoms, mostly biogenic detritus (including fragments of foraminiferal tests), coccoliths and mineral particles (Morozov et al. 2010).

3 Geology

The evolution of the eastern Brazilian rift system includes three main syn-rift phases that took place between 150–120 Ma (Ma = million years ago). It was characterized by the formation of the great depression in the Gondwana supercontinent and intense basaltic volcanism. The rift phases were followed by an evaporitic phase under restricted oceanic conditions (Chang et al. 1992). In terms of the general characteristics of the basement and distribution of the main structural features, the southern part of the Brazil Margin and its adjacent abyssal plains falls into four domains (Bassetto et al. 2000). Domain I covers a large segment to the south of latitude 32°S near the

Porto Alegre Lineament, including the continental slope, continental rise and abyssal plain. It is characterized by relatively thin igneous oceanic crust with a thickness of between 5 and 6.5 km (Bassetto et al. 2000). Domain II lies between the Porto Alegre Lineament and the Florianopolis Fracture Zone. Recent studies have demonstrated that the basement in the area of the Santa Catarina Plateau relating to this domain is mostly constituted of continental crust (Jeck et al. 2019). The crustal thickness varies from 4.0 to 10.0 km (Bassetto et al. 2000). Domain III covers the area of the São Paulo Plateau and is part constituted of extended continental crust and part of probably transitional crust. It is characterized by the highly irregular relief of the basement, resulting from numerous volcanic constructions (Bassetto et al. 2000). Domain IV corresponds to the area located north of the Florianopolis Fracture Zone and east of the São Paulo Plateau. This domain forms this monograph's main area of interest. The entire domain is on oceanic crust, beyond the geographic boundaries of the continental slope and rise. It demonstrates a relative morpho-structural homogeneity and is characterized by a crustal thickness of 5.5–6.5 km (Bassetto et al. 2000). The oceanic basement in the main focus area is of Cenomanian–Coniacian age (95–80 Ma, after Müller et al. 1997).

4 Sedimentology

In this part of the western South Atlantic, the modern regime of oceanic sedimentation started during the Aptian/Albian (113 Ma, after Chang et al. 1992). Late Cretaceous–Eocene sedimentation was probably dominated by hemipelagic settling and turbidity currents. The opening of the Drake and Tasman oceanic seaways in the southern hemisphere during the Eocene–Oligocene (e.g., Allen and Armstrong 2008) caused dramatic changes in ocean circulation. Since that time, deposition has been controlled by the interaction between along-slope processes (geostrophic (contour) currents) and gravity-driven down-slope processes (gravity flows and mass-wasting from the continental slope and seamounts). Global climate changes, along with tectonic movements, have resulted in an increased production of NADW and AABW, which caused episodes of drastic erosion and the formation of several major regional unconformities. These unconformities represent timelines of the period from the Eocene–Oligocene to the late Pliocene (Gamboa et al. 1983; Viana et al. 2003; Lima et al. 2009).

The sediment cover in the Rio Grande Gap area and the southern Brazil Basin shows four seismic sequences with various reflection configuration patterns, divided by distinct unconformities (Gamboa et al. 1983). The main erosional event (hiatus from the Early Eocene to Middle Oligocene) is marked by the pronounced seismic reflector "A". This oldest prominent unconformity separates the lower sequence of acoustically transparent sediments from three overlying sequences of highly stratified sediments. This change in acoustic structure probably corresponds to the change in the sedimentation mode due to the onset of AABW flow from the Argentine Basin to the Brazil Basin through the Vema Channel. The other two unconformities, which separate three upper stratified sequences, are erosional surfaces resulting from significant changes to the bottom bottom-water circulation. These unconformities are of the Late Oligocene and Late Miocene in age (Barker et al. 1983a; Gamboa et al. 1983; Camboa and Rabinowitz 1984).

Sediment thickness in the abyssal part of the Southern Brazil Basin exceeds 1km. According to results from the Deep-Sea Drilling Project (DSDP) at Site 515 (water depth 4,252 m), located north-west of the study area (Fig. 1a), the upper 180 m of the recovered section is composed of terrigenous mud and clay with occasional nannofossil-rich layers and some foraminifer-rich layers. The Pleistocene/Pliocene boundary is found at a depth of about 40m below sea floor (mbsf). There are two major hiatuses in the depositional record. The first is revealed in the Late Miocene and another extends from 50 to 32 Ma and spans the Eocene/Oligocene boundary (Barker et al. 1983c). Data collected from DSDP Sites 516–518 (water depths 1,313 m, 2,963 m and 3,944 m, respectively), located on the summit and western flank of the Rio Grande Rise southwest of the study area (Fig. 1a), suggest a thickness of Quaternary sediments of about 8–27 m. Recovered sediments are represented by calcareous oozes (foraminiferal-nannofossil ooze, foraminiferal sand and nannofossil ooze) down to 193 m. Below this depth at Site 516, a recovered succession (of about 800 m in thickness) consisting of nannofossil and foraminiferal-nannofossil chalks, foraminiferal limestones, limestones and marly limestones covers a timespan from the Paleocene to the Miocene (Barash et al., 1983; Barker et al. 1983b,d; Barash 1988). In the sediment sequence recovered from Site 518, a hiatus spans the Late Miocene–Early Pliocene interval (from 5.5 to 4.0 Ma). Another major hiatus is revealed in the Middle Miocene (Barker et al. 1983d).

The south-east Brazilian Margin presents a sedimentary prism extending from the continental shelf into the deep ocean, characterized by gradually decreasing thickness. This characteristic is masked by the seafloor's morphology and sedimentary processes, especially those related to bottom currents (Jeck et al. 2019).

Previous studies revealed several prominent contourite features in the region between 35 and 20°S (Fig. 1a): the Pelotas Drift, the Chui Drift (Jeck et al. 2019), the Ewing Drift (also referred to as the Rio Grande Drift) (Ewing and Lonardi 1971; Flood and Shor 1988; Klaus and Ledbetter 1988; Flood et al. 1993), the Vema contourite fan (Mézerais et al. 1993), the Ioffe Drift (Ivanova et al. 2016, 2020), the Columbia Channel depositional system (Faugères et al. 2002b; Lima et al. 2009), the Santos contourite depositional system (Duarte and Viana 2007; De Almeida et al. 2015; Toledo et al. 2016) and plastered drifts on the São Paulo Plateau escarpment (Borisov et al. 2013; Ovsepyan and Ivanova 2019).

The Santos contourite depositional system in the Santos Basin (SE Brazilian Margin) developed from the Oligocene to Recent under the influence of strong along-slope currents. It is located at a water depth range of 500–3,000 m and is composed of several contourite features (Duarte and Viana 2007). A field of confined drifts covered by sediment waves on the Santa Catarina Plateau is known as the Santa Catarina Drift (Jeck et al. 2019). The Pelotas Drift is an elongated depositional body attached to the southern border of the Santa Catarina Plateau, smoothing its scarp. The drift characterized by slope instabilities such as normal faults and mass-transport deposits remains understudied (Jeck et al. 2019). The Ewing elongated arc-shaped mounded drift, partly covered by sediment waves, extends in an east to south-east direction from the South American continental rise into the abyssal plain toward the Zapiola Drift (e.g., Hernández-Molina et al. 2010). South of the Ewing Drift, there is a smaller elongated mounded depositional body known as the Chui Drift, whose crest extends NE–SW (Jeck et al. 2019).

The crest orientation of the Ewing, Pelotas and Chui Drifts is consistent with the general pathway predicted for a return flow of abyssal water (Flood and Shor 1988; Hernández-Molina et al. 2010). Sediment wave fields were revealed in the northern part of the Santa Catarina Plateau (Borisov et al. 2020) and on the Rio Grande Rise (Lisniowski et al. 2017; Levchenko et al. 2020). The Vema contourite fan (drift) is located at a depth of approximately 4,500 m near the northern exit from the Vema Channel. It has a length of 150 km and a width of 60–120 km, increasing downstream toward the northeast (Mézerais et al. 1993). The Neogene evolution of the body is characterized by alternating periods of deposition (low bottom-current intensity) and erosion (high bottom-current intensity), the latter commonly underlined by truncation surfaces (Mézerais et al. 1993; Faugères et al. 2002a). The Columbia Channel system located south of the Vitoria–Trindade seamounts is considered to be a mixed turbidite–contourite system. It is composed of the Columbia Channel itself and two depocenters located north and south of the channel. The system is 500 km in length and 200–350 km wide. It overlies an Upper Cretaceous–Paleogene sedimentary substratum and was actively developed throughout the Oligocene–Neogene. The interplay between bottom currents and turbidity flows resulted in the formation of depositional features typical of a mixed turbidite–contourite system (Faugères et al. 2002b; Lima et al. 2009).

References

Allen MB, Armstrong HA (2008) Arabia-Eurasia collision and the forcing of mid-Cenozoic global cooling. Palaeogeogr Palaeoclimatol Palaeoecol 265:52–58. https://doi.org/10.1016/j.palaeo.2008.04.021

Barash MS (1988) Quaternary paleooceanology of the Atlantic Ocean. Nauka (in Russian), Moscow

Barash MS, Oskina NS, Bylum NS (1983) Quaternary biostratigraphy and surface paleotemperatures based on planktonic foraminifers. In: Barker PF, Johnson DA, Carlson RL (eds) Initial Reports of the Deep Sea Drilling Project, 72. U.S. Government Printing Office, Washington, pp 849–869

Barker PF, Buffler RT, Gamboa LA (1983a) A seismic reflection study of the Rio Grande Rise. In: Barker PF, Johnson DA, Carlson RL (eds) Initial Reports of the Deep Sea Drilling Project, 72. U.S. Government Printing Office, Washington

Barker PF, Johnson DA, Carlson RL et al (1983) Site 516: Rio Grande Rise. In: Barker PF, Johnson DA, Carlson RL (eds) Initial Reports of the Deep Sea Drilling Project, 72. Government Printing Office, Washington, U.S, pp 155–338

Barker PF, Johnson DA, Carlson RL et al (1983) Site 515: Brazil Basin. In: Barker PF, Johnson DA, Carlson RL (eds) Initial Reports of the Deep Sea Drilling Project, 72. Government Printing Office, Washington, U.S, pp 54–154

Barker PF, Johnson DA, Carlson RL et al (1983) Site 518: West Flank, Rio Grande Rise. In: Barker PF, Johnson DA, Carlson RL (eds) Initial Reports of the Deep Sea Drilling Project, 72. Government Printing Office, Washington, U.S, pp 357–380

Bassetto M, Alkmim FF, Szatmari P, Mohriak WU (2000) The oceanic segment of the Southern Brazilian margin: Morpho-structural domains and their tectonic significance. Geophys Monogr Ser 115:235–299. https://doi.org/10.1029/GM115p0235

Biscaye PE, Eittreim SL (1977) Suspended particulate loads and transports in the nepheloid layer of the abyssal Atlantic Ocean. Mar Geol 23:155–172. https://doi.org/10.1016/0025-3227(77)90087-1

Borisov D, Frey D, Levchenko O (2020) Sediment waves on the Santa Catarina Plateau (western South Atlantic). J South Am Earth Sci 102:102698. https://doi.org/10.1016/j.jsames.2020.102698

Borisov DG, Murdmaa IO, Ivanova EV et al (2013) Contourite systems in the region of the southern São Paulo Plateau escarpment, South Atlantic. Oceanology 53:460–471. https://doi.org/10.1134/S0001437013040012

Bueno GV, Machado DLJ, Oliveira JAB, Marques EJJ (2004) A influência do Lineamento Capricrrnio na evoluqao tectono-sedimentar da Bacia de Santos. In: *Congresso Brasileiro de Geologia,* 52, Araxá. Annals, Sociedade Brasileira de Geologia, Simposium 28-Petróleo: Geologia e Exploração. p T773

Camboa LAP, Rabinowitz PD (1984) The evolution of the Rio Grande Rise in the southwest Atlantic Ocean. Mar Geol 58:35–58. https://doi.org/10.1016/0025-3227(84)90115-4

Campos CWM, Ponte FC, Miura K (1974) Geology of the Brazilian continental margin. In: The geology of continental margins. Springer, Berlin, pp 447–461

Chang HK, Kowsmann RO, Figueiredo AMF, Bender AA (1992) Tectonics and stratigraphy of the East Brazil Rift system: an overview. Tectonophysics 213:97–138. https://doi.org/10.1016/0040-1951(92)90253-3

De Almeida FK, De Mello RM, Costa KB, Toledo FAL (2015) The response of deep-water benthic foraminiferal assemblages to changes in paleoproductivity during the Pleistocene (last 769.2kyr), western South Atlantic Ocean. Palaeogeogr Palaeoclimatol Palaeoecol 440:201–212. https://doi.org/10.1016/j.palaeo.2015.09.005

Dianskii NA, Bagno AV, Zalesny VB (2002) Sigma-model for global ocean circulation and its sensitivity to variations in wind friction stresses. Izv Akad Nauk Fiz Atmos I OKEANA 38:537–556

Diretoria de Hidrografia e Navegação, Marinha do Brasil (2017) Fisiografia da margem brasileira - Modelo Digital do Terreno

Duarte CSL, Viana AR (2007) Santos drift system: stratigraphic organization and implications for late Cenozoic palaeocirculation in the Santos Basin, SW Atlantic Ocean. Geol Soc London, Spec Publ 276:171–198. https://doi.org/10.1144/GSL.SP.2007.276.01.09

Duncan RA, Richards MA (1991) Hotspots, mantle plumes, flood basalts, and true polar wander. Rev Geophys 29:31. https://doi.org/10.1029/90RG02372

Ewing M, Lonardi AG (1971) Sediment transport and distribution in the Argentine Basin. 5. Sedimentary structure of the Argentine margin, basin, and related provinces. Phys Chem Earth 8:125–156. https://doi.org/10.1016/0079-1946(71)90017-6

Faugères J-C, Zaragosi S, Mézerais ML, Massé L (2002) The Vema contourite fan in the South Brazilian Basin. Geol Soc London Mem 22:209–222. https://doi.org/10.1144/GSL.MEM.2002.022.01.15

Faugères J-C, Lima AF, Massé L, Zaragosi S (2002) The Columbia channel-levee system: a fan drift in the southern Brazil Basin. Geol Soc London Mem 22:223–238. https://doi.org/10.1144/GSL.MEM.2002.022.01.16

Flood RD, Shor AN (1988) Mud waves in the Argentine Basin and their relationship to regional bottom circulation patterns. Deep Sea Res Part A Oceanogr Res Pap 35:943–971. https://doi.org/10.1016/0198-0149(88)90070-2

Flood RD, Shor AN, Manley PL (1993) Morphology of abyssal mudwaves at project MUDWAVES sites in the Argentine Basin. Deep Res Part II 40:859–888. https://doi.org/10.1016/0967-0645(93)90038-O

Fodor RV, Hanan BB (2000) Geochemical evidence for the Trindade hotspot trace: Columbia seamount ankaramite. Lithos 51:293–304. https://doi.org/10.1016/S0024-4937(00)00002-5

Fontana RL (1996) Geotectônica e sismoestratigrafia da Bacia de Pelotas e Plataforma de Florianópolis. Dr Dissertation, Univ Fed do Rio Gd do Sul, Porto Alegre

Frey DI, Fomin VV, Tarakanov RY et al (2018) Bottom water flows in the Vema Channel and over the Santos Plateau: Based on the field and numerical experiments. In: Velarde MG, Tarakanov RY, Marchenko AV (eds) The ocean in motion: circulation, waves, polar oceanography. Springer International Publishing, Cham, pp 475–485

Frey DI, Morozov EG, Fomin VV et al (2019) Regional modeling of Antarctic bottom water flows in the key passages of the Atlantic. J Geophys Res Ocean 124:8414–8428. https://doi.org/10.1029/2019JC015315

Gamboa LA, Buffler RT, Barker PF (1983) Seismic stratigraphy and geologic history of the Rio Grande Gap and Southern Brazil Basin. In: Barker PF, Johnson DA, Carlson RL (eds) Initial reports of the deep-sea drilling project, 72. Government Printing Office, Washington, U.S, pp 481–496

Gonthier E, Faugères JC, Viana A et al (2003) Upper Quaternary deposits on the São Tomé deep-sea channel levee system (South Brazilian Basin): major turbidite versus contourite processes. Mar Geol 199:159–180. https://doi.org/10.1016/S0025-3227(03)00128-2

Hernández-Molina FJ, Paterlini M, Somoza L et al (2010) Giant mounded drifts in the Argentine continental margin: origins, and global implications for the history of thermohaline circulation. Mar Pet Geol 27:1508–1530. https://doi.org/10.1016/j.marpetgeo.2010.04.003

Hernández-Molina FJ, Soto M, Piola AR et al (2016) A contourite depositional system along the Uruguayan continental margin: sedimentary, oceanographic and paleoceanographic implications. Mar Geol 378:333–349. https://doi.org/10.1016/j.margeo.2015.10.008

Ivanova E, Borisov D, Dmitrenko O, Murdmaa I (2020) Hiatuses in the late Pliocene-Pleistocene stratigraphy of the Ioffe calcareous contourite drift, western South Atlantic. Mar Pet Geol 111:624–637. https://doi.org/10.1016/j.marpetgeo.2019.08.031

Ivanova E, Murdmaa I, Borisov D et al (2016) Late Pliocene–Pleistocene stratigraphy and history of formation of the Ioffe calcareous contourite drift. Western South Atlantic. Mar Geol 372:17–30. https://doi.org/10.1016/j.margeo.2015.12.002

Jackett DR, McDougall TJ (1997) A neutral density variable for the world's oceans. J Phys Oceanogr 27:237–263. https://doi.org/10.1175/1520-0485(1997)0272.0.CO;2

Jeck IK, Alberoni AAL, Torres LC, Zalán PV (2019) The Santa Catarina Plateau and the nature of its basement. Geo-Marine Lett. https://doi.org/10.1007/s00367-019-00585-z

Johnson DA (1984) The Vema Channel: physiography, structure, and sediment–current interactions. Mar Geol 58:1–34.https://doi.org/10.1016/0025-3227(84)90114-2

Jullion L, Heywood KJ, Naveira Garabato AC, Stevens DP (2010) Circulation and water mass modification in the Brazil-Malvinas confluence. J Phys Oceanogr 40:845–864. https://doi.org/10.1175/2009JPO4174.1

Klaus A, Ledbetter MT (1988) Deep-sea sedimentary processes in the Argentine Basin revealed by high-resolution seismic records (3.5 kHz echograms). Deep Sea Res Part A Oceanogr Res Pap 35:899–917. https://doi.org/10.1016/0198-0149(88)90067-2

Kowsmann R, Leyden O (1977) Marine seismic investigations, Southern Brazil Margin. Am Assoc Pet Geol Bull 61:546–557. https://doi.org/10.1306/C1EA3D84-16C9-11D7-8645000102C1865D

Larqué L, Maamaatuaiahutapu K, Garçon V (1997) On the intermediate and deep water flows in the South Atlantic Ocean. J Geophys Res Ocean 102:12425–12440. https://doi.org/10.1029/97JC00629

Levchenko OV, Borisov DG, Libina NV (2020) Contourites and gravitites on the Rio Grande Rise, Southwest Atlantic Ocean (Seismoacoustic Data). Lithol Miner Resour 55:165–176. https://doi.org/10.1134/S0024490220030037
Levitus S, Locarnini RA, Boyer TP et al (2010) World ocean Atlas 2009
Lima AF, Faugeres JC, Mahiques M (2009) The Oligocene-Neogene deep-sea Columbia channel system in the South Brazilian Basin: seismic stratigraphy and environmental changes. Mar Geol 266:18–41. https://doi.org/10.1016/j.margeo.2009.07.009
Lisniowski MA, Harlamov V, Frazao EP et al (2017) Sediment waves on the Rio Grande Rise. In: 2017 IEEE/OES acoustics in underwater geosciences symposium (RIO Acoustics). IEEE, pp 1–5
Mascle J (1976) Atlantic-type continental margins: distinction of two basic structural types. Acad Bras Ciências 48:191–197
Mézerais ML, Faugères JC, Figueiredo AG, Massé L (1993) Contour current accumulation off the Vema Channel mouth, southern Brazil Basin: pattern of a "contourite fan." Sediment Geol 82:173–187. https://doi.org/10.1016/0037-0738(93)90120-T
Morozov EG, Demidov AN, Tarakanov RY, Zenk W (2010) Abyssal channels in the Atlantic Ocean. Springer, Netherlands, Dordrecht
Müller RD, Roest WR, Royer J-Y et al (1997) Digital isochrons of the world's ocean floor. J Geophys Res Solid Earth 102:3211–3214. https://doi.org/10.1029/96JB01781
Orsi AH, Johnson GC, Bullister JL (1999) Circulation, mixing, and production of Antarctic bottom water. Prog Oceanogr 43:55–109. https://doi.org/10.1016/S0079-6611(99)00004-X
Ovsepyan EA, Ivanova EV (2019) Glacial–interglacial interplay of southern- and northern-origin deep waters in the São Paulo Plateau-Vema Channel area of the western South Atlantic. Palaeogeogr Palaeoclimatol Palaeoecol 514:349–360. https://doi.org/10.1016/j.palaeo.2018.10.031
Pätzold J, Heidland K, Zenk W, Siedler G (1996) On the bathymetry of the Hunter Channel. In: The South Atlantic. Springer, Berlin, pp 355–361
Piola AR, Matano RP (2001) Brazil and Falklands (malvinas) Currents. In: Steele JH, Thorpe SA, Turekian KK (eds) Encyclopedia of ocean sciences, Academic P. Elsevier, London, pp 340–349
Preu B, Hernández-Molina FJ, Violante R et al (2013) Morphosedimentary and hydrographic features of the northern Argentine margin: the interplay between erosive, depositional and gravitational processes and its conceptual implications. Deep Res Part I Oceanogr Res Pap 75:157–174. https://doi.org/10.1016/j.dsr.2012.12.013
Reid JL, Nowlin WD, Patzert WC (1977) On the characteristics and circulation of the Southwestern Atlantic Ocean. J Phys Oceanogr 7:62–91
Schmid C, Siedler G, Zenk W (2000) Dynamics of intermediate water circulation in the subtropical South Atlantic. J Phys Oceanogr 30:3191–3211. https://doi.org/10.1175/1520-0485(2000)030%3c3191:DOIWCI%3e2.0.CO;2
Skolotnev SG, Peyve AA, Turko NN (2010) New data on the structure of the Vitoria-Trindade seamount chain (western Brazil basin, South Atlantic). Dokl Earth Sci 431:435–440. https://doi.org/10.1134/S1028334X10040057
Tarakanov RY, Morozov EG (2015) Flow of Antarctic bottom water at the output of the Vema Channel. Oceanology 55:153–161. https://doi.org/10.1134/S0001437015010166
Toledo FAL, Quadros JP, Camillo E et al (2016) Plankton biochronology for the last 772,000 years from the western South Atlantic Ocean. Mar Micropaleontol 127:50–62. https://doi.org/10.1016/j.marmicro.2016.07.002
Viana A, Figueiredo AG, Faugères JC et al (2003) The São Tomé deep-sea turbidite system (Southern Brazil Basin): Cenozoic seismic stratigraphy and sedimentary processes. Am Assoc Pet Geol Bull 87:873–894. https://doi.org/10.1306/12100201048

Regional Stratigraphic Frameworks Based on Calcareous Microfossils

Elena Ivanova and Olga Dmitrenko

Abstract In this chapter we consider the previously published stratigraphic schemes and datum levels of planktic foraminiferal and nannofossil index species from the Atlantic. The stratigraphic frameworks developed for the Ioffe Drift in this study covers the last 3.8 Ma recovered from sediment cores collected from the area. The Pliocene/Pleistocene boundary is placed at about 2.59 Ma, according to the generally accepted International Chronostratigraphic Chart (Cohen et al. (2013) The ICS international chronostratigraphic chart. Episodes 36; Cohen et al. (2020) The ICS international chronostratigraphic chart. Episodes 36). The ages of zonal boundaries are generally defined by the index species datum levels, that is, first or last occurrences, mostly adopted from recent publications.

1 Major Principles of Developing a Stratigraphic Framework for the Study Area

As the sediments of the Ioffe Drift are generally represented by foraminiferal and nanno-foraminiferal oozes, these microfossils' zonation was applied to ascertain the age of the recovered core sections (Ivanova et al. 2016, 2020). It should be underlined that, in the case of the Ioffe Drift, classical biostratigraphy provides the most reliable stratigraphic and age control to identify the major hiatuses and shed light on the timing of the major paleoceanographic events during its formation.

Foraminiferal and nannofossil stratigraphic frameworks, mostly from the Central and South Atlantic yet also from other locations as well as the species' datum levels (first and/or last occurrences, FO and/or LO, respectively) were used to identify biostratigraphic zones and their boundaries (Blow 1969; Bolli and Premoli Silva 1973; Martini 1971; Gartner 1977; Berggren et al. 1983a, b, 1995a, b; Kennett

E. Ivanova (✉) · O. Dmitrenko
Shirshov Institute of Oceanology, Russian Academy of Sciences, Moscow, Russia
e-mail: e_v_ivanova@ocean.ru

O. Dmitrenko
e-mail: senidol@yandex.ru

I. Murdmaa and E. Ivanova (eds.), *The Ioffe Drift*, Springer Geology,
https://doi.org/10.1007/978-3-030-82871-4_3

and Srinivasan 1983; Bolli and Saunders 1985; Chaisson and Pearson 1997; Okada 2000; Bylinskaya and Golovina 2004). Priority was given to publications dealing with the subtropical South Atlantic and the most detailed schemes. Regarding the datum levels, these were mostly adopted from recent publications (Wade et al. 2011; Backman et al. 2012; Browning et al. 2017; Bergen et al. 2019). Accordingly, a revised stratigraphy of the previously published version for three of the six cores from the Ioffe Drift (Ivanova et al. 2016, 2020) is presented in Chap. 7. In particular, the previously adopted in our publications (Ivanova et al. 2016, 2020) foraminiferal zonation of the nearby Rio Grande Rise for the Pleistocene (Barash et al. 1983; Barash 1988) and Pliocene (Berggren et al. 1983a, b, 1995a, b) is now supplemented by data from Wade et al. (2011) on tropical Pliocene zonation and datum levels. Likewise, the nannofossil datum levels and zonations (Martini 1971; Gartner 1977; Berggren et al. 1983a; Dmitrenko 1987; Okada 2000; Bylinskaya and Golovina 2004) that were used to stratify our first core AI-2436 (Ivanova et al. 2016) have been supplemented by more recent data from the western tropical Atlantic (Backman et al. 2012) and the Gulf of Mexico (Browning et al. 2017; Bergen et al. 2019), notably for the Pliocene. The correlations of the above-mentioned biostratigraphic frames and datum levels applied are shown in Figs. 1 and 2, respectively.

It worth noting that we consider the Pliocene/Pleistocene boundary dated to about 2.59 Ma, in line with the generally accepted International Chronostratigraphic Chart (Cohen et al. 2013, 2020), while in the above-mentioned publications it was placed at 1.88–1.9 Ma. Thus, we include the Gelasian stage in the Pleistocene and regard the Pliocene as covering the Piacenzian and Zanclean (Cohen et al. 2013, 2020). Unfortunately, in contrast to the earlier definition of the Pliocene/Pleistocene boundary as at the FO of the rather common subtropical-tropical foraminiferal species *Globorotalia truncatulinoides* at about 1.9 Ma (Figs. 1 and 2), its present position on the chronostratigraphic scale at 2.59 Ma is unsupported by any marine microfossil datum level. Moreover, the pronounced evolutionary disappearance of discoasters close to the FO level of *G. truncatulinoides* can no longer be used as the criterion to define the Pliocene/Pleistocene boundary. This creates difficulties in developing the regional microfossil-based stratigraphy, notably in the study area as the boundary cannot be precisely determined in any sediment core.

2 Planktic Foraminiferal Zonation

For the Quaternary (2.59–0 Ma, according to Cohen et al. 2013, 2020) and Upper Pliocene sections, the following foraminiferal zones and subzones were previously identified in DSDP Site 516 (and partly in Site 518) from the Rio Grande Rise (Fig. 1; Barash et al. 1983; Barash 1988) for the last 1.9 Ma and the older interval (Berggren et al. 1983a, b, 1995a, b).

The uppermost *Globigerinoides ruber* (pink) subzone of zone N23 (Blow 1969) corresponds to the index species occurrence since 0.28 Ma (Barash et al. 1983; Barash 1988). This subzone is specific to the area while at other locations, like the tropical

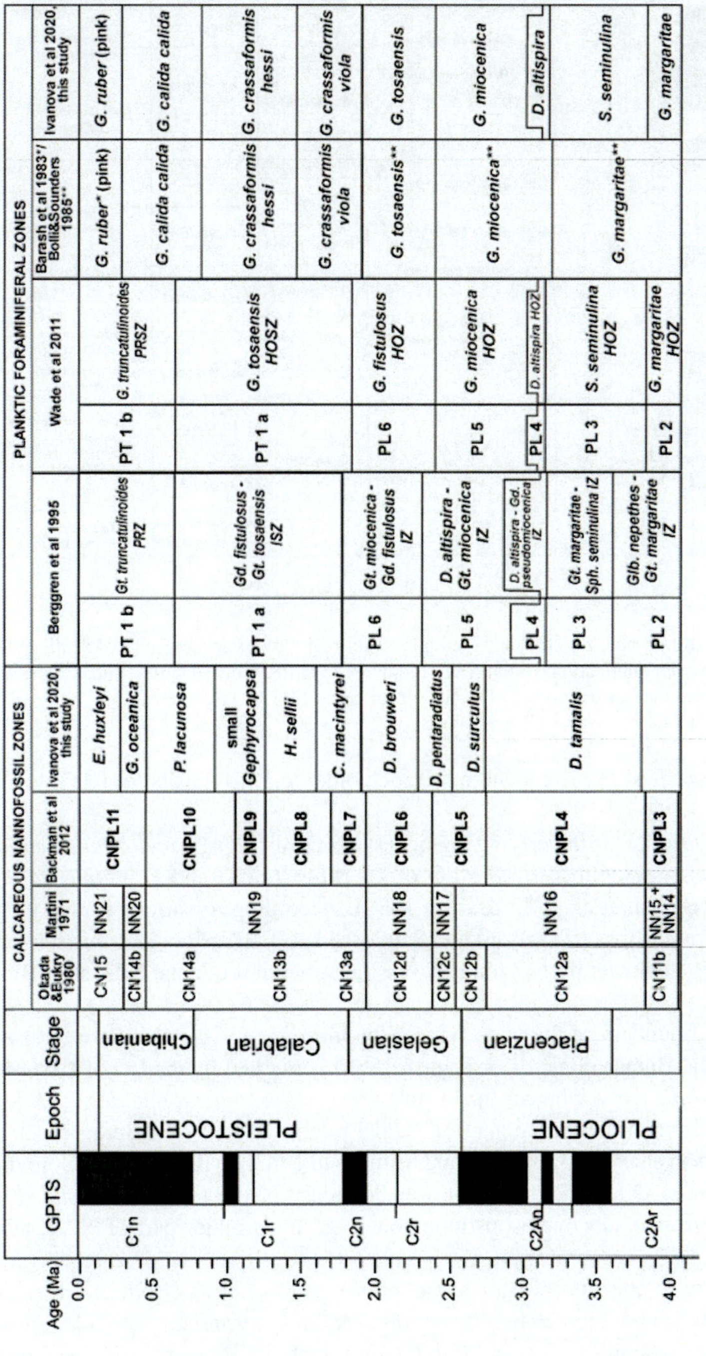

Fig. 1 Comparison of upper Pliocene–Pleistocene calcareous and planktic foraminiferal biostratigraphy for the Tropical and South Atlantic from several studies. GPTS–Geomagnetic Polarity Time Scale (Lourens et al. 2004), PRZ–partial-range zone, PRSZ–partial-range subzone, ISZ–interval subzone, IZ–interval zone, HOZ–highest occurrence zone, HOSZ–highest occurrence subzone. *Barash et al. 1983; **Bolli and Saunders 1985

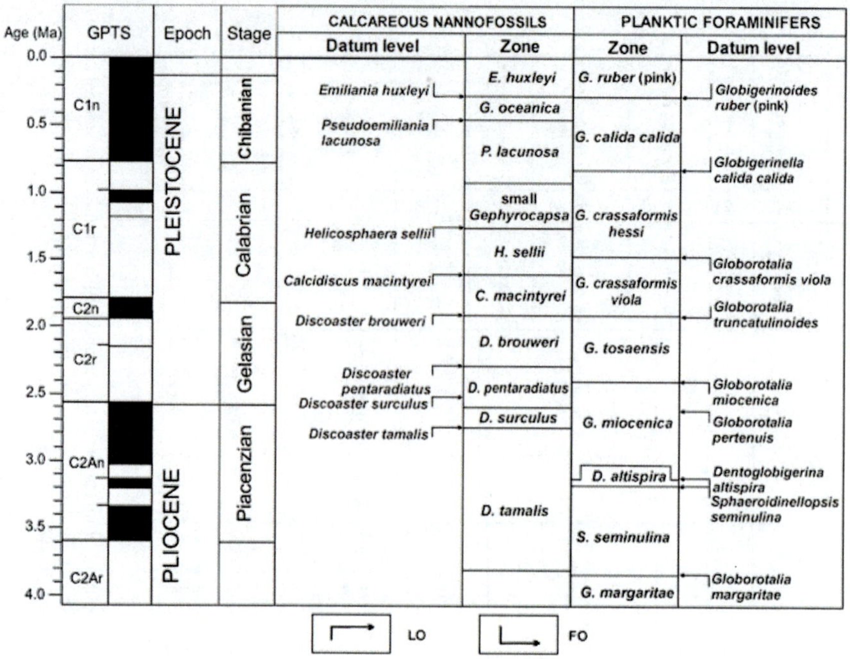

Fig 2 The biostratigraphic zonation and datum levels of major index species in the Atlantic in this study are from earlier publications (see In sects. "Planktic Foraminiferal Zonation" and "Calcareous Nannofossil Zonation" for details). (1) FO–first occurrence; (2) LO–last occurrence

Atlantic, the species' FO is known to be much older (e.g., Bylinskaya and Golovina 2004).

The boundaries of the underlying regional subzone of *Globigerinella calida calida* (0.81–0.28 Ma) are defined by the FO of *G. ruber* (pink) and *G. calida calida*, respectively (Barash et al. 1983; Barash 1988). According to various studies (e.g., Bolli and Premoli Silva 1973; Bolli and Saunders 1985; Barash 1988), this subzone corresponds to the lower part of zone N23 or the upper part of zone N22, after Blow (1969).

The lower boundary of the underlying *Globorotalia crassaformis hessi* subzone (1.47–0.81 Ma, Barash et al. 1983; Barash 1988) is marked by the LO of *Globorotalia crassaformis viola,* whereas the FO of *Globorotalia truncatulinoides* (~1.9 Ma, according to several authors, e.g., Bolli and Premoli Silva 1973; Berggren et al. 1983a, b; Barash et al. 1983; Kennett and Srinivasan 1983; Bylinskaya and Golovina 2004; Wade et al. 2011) defines the lower boundary of the *G. crassaformis viola* subzone. These two subzones constitute zone N22 (or a major part of N22), after Blow (1969).

The lowermost Pleistocene subzone *Globorotalia tosaensis* (Bolli and Premoli Silva 1973; Bolli and Saunders 1985) or subzone *Globigerinoides fistulosus* (Wade et al. 2011), corresponding to zone PL6 (Berggren et al. 1995a, b), was previously attributed to the Upper Pliocene, notably at DSDP Site 516 (Berggren et al. 1983a); however, according to the current definition of the Pliocene/Pleistocene boundary at 2.59 Ma (Cohen et al. 2013, 2020), this subzone is included in the Pleistocene section. The lower boundary of the *Globorotalia tosaensis* subzone is inferred from the disappearance of some Neogene species, most importantly by the LO of *Globorotalia miocenica* at 2.3 or 2.39 Ma, according to Chaisson and Pearson (1997) and Wade et al. (2011), respectively. Previously, researchers had suggested the LO of *G. miocenica* in the tropical Atlantic was ~2.4 Ma (Bolli and Saunders 1985; Bylinskaya and Golovina 2004).

It should be noted that the subzones of *Globigerinella calida calida* and *Globorotalia crassaformis hessi* cannot be identified in some of our core sections, as discussed in Sect. "Methods". *Globorotalia tosaensis* and *Globigerinoides fistulosus* occur rather rarely in the material studied and cannot be used to identify subzones of the same name. The LO of *Globorotalia pertenuis* at 2.6 Ma (Fig. 2; Wade et al. 2011), representing the datum closest to the Pliocene/Pleistocene boundary, is not very useful due to the rare occurrence of this tropical species in the subtropics, notably in the study area. The LO of another important species, *Globorotalia multicamerata,* which is more common in our material, is about 3 Ma (Wade et al. 2011). Previous studies estimated this at ~2.5 Ma (Berggren et al. 1983a, b), 2.91 Ma (Leonard et al. 1983) or 3.09 Ma (Chaisson and Pearson 1997); that is, in the lowermost part of the *Globorotalia miocenica* zone, and anyway above the LO of *Dentoglobigerina altispira.*

As authors variously use the term "zone" or "subzone" to refer to the same foraminiferal (and nannoplankton) units, for convenience and consistency throughout the text and in the figures, including Figs. 1 and 2, all the above-mentioned subzones are referred to as "zones".

In the Pliocene (and lowermost Pleistocene) section (~4–2.39 Ma, in the cores studied), three foraminiferal zones were identified using the datum levels from Wade et al. (2011) for the tropical Atlantic. The zone of *Globorotalia miocenica* (3.13–2.39 Ma), or PL5 (Berggren et al. 1995a, b), is defined as between the LO of the zonal index species and that of *Dentoglobigerina altispira* (Figs. 1 and 2). The lower boundary of the shortest zone of *Dentoglobigerina altispira* (3.16–3.13 Ma), or PL4, corresponds to the LO of *Sphaeroidinellopsis seminulina,* the index species of the underlying zone of the same name (3.84–3.16 Ma), or PL3. The lower zonal boundary cannot be robustly documented in any of the core sections studied due to the absence of the index species of PL2, *Globorotalia margaritae.*

In our cores from the Ioffe Drift, the most reliable foraminiferal datum levels seem to be the base of the zones of *Globigerinoides ruber* (pink) (0.28 Ma) and *Globorotalia crassaformis viola* (1.9 Ma), as well as the boundaries of the *Dentoglobigerina altispira* zone (3.16 and 3.13 Ma) (Fig. 2).

3 Calcareous Nannofossil Zonation

The following detailed Pliocene–Quaternary nannofossil zonation according to Gartner (1977) was applied to stratify the four core sections from the Ioffe Drift, as follows, with the species datum level from the same study and from Martini (1971), with revisions by later studies in the Rio Grande Rise, Northern and Topical Atlantic (e.g., Berggren et al. 1983a; Okada 2000; Bylinskaya and Golovina 2004; Backman et al. 2012; Browning et al. 2017; Bergen et al. 2019).

The uppermost zone of *Emiliania huxleyi* (0.27–0 Ma), including the *E. huxleyi* Acme zone (0.07–0 Ma), corresponds to the occurrence of the index species of the same name from 0.2 to 0.29 Ma (Martini 1971; Gartner 1977; Bukry 1978; Okada and Bukry 1980; Backman et al. 2012) and maximum abundance, respectively; however, Backman et al. (2012) notes the time-transgressive character of the species Acme zone, following Thierstein et al. (1977).

The *Gephyrocapsa oceanica* zone (0.44–0.27 Ma) is defined as between the FO of *Emiliania huxleyi* and the LO of *Pseudoemiliania lacunosa.* Although *Gephyrocapsa oceanica* (and *G. caribbeanica*) commonly occurs in Quaternary sediments within the interval 1.9–0.27 Ma, this zone is not considered in some recent studies (e.g., Backman et al. 2012), yet it was previously identified in the nearby Rio Grande Rise (Dmitrenko 1987) and in some cores from our study (see Chap. 7).

The age of the underlying *Pseudoemiliania lacunosa* zone (0.92–0.44 Ma, according to Gartner 1977), varies significantly according to different studies: 0.6–1.8 Ma (Martini 1971), 0.46–1.95 Ma (Okada 2000) and 0.43–1.06 Ma (Backman et al. 2012), because of the prolonged occurrence of the species of this name and the difficulties in identifying the underlying zone of the small *Gephyrocapsa* (1.25–0.92 Ma in this study; 1.22–0.92 Ma, after Gartner 1977); and 1.25–1.06 Ma, after Backman et al. 2012). The tiny coccoliths of the nominated taxa commonly cannot be found in sediment sections, notably in the Ioffe Drift. In Site 516 of the Rio Grande Rise, the upper zonal boundary is marked by a pronounced decrease in the abundance and diversity of small *Gephyrocapsa.*

The boundary between the zones of small *Gephyrocapsa* and *Helicosphaera sellii* is defined by Gartner (1977) by the LO of the latter taxon, dated to 1.25 Ma by Bergen et al. (2019). The *Helicosphaera sellii* zone (1.6–1.25 Ma, in this study; 1.51–1.22 Ma, after Gartner 1977) can be identified only in some core sections, including one from the Ioffe Drift (see Chap. 7), due to the rather rare occurrence and/or poor preservation of the self-titled species in the deep-sea sediments from this time span. Therefore, it is unsurprising that this zone is not considered by several authors (e.g., Martini 1971; Backman et al. 2012).

The underlying *Calcidiscus macintyrei* zone (1.9–1.6 Ma, in this study 1.65–1.51 Ma, after Gartner 1977; and 1.93–1.71 Ma, after Backman et al. 2012) is also included in the zone of *Pseudoemiliania lacunosa* by Martini (1971), like the zone of small *Gephyrocapsa;* however, it is well defined in two of our cores and in DSDP Site 516 (Dmitrenko 1987). Gartner (1977) placed the lower boundary of the zone of *Calcidiscus macintyrei*; that is, its contact with the underlying *Discoaster brouweri* zone, at 1.65 Ma, although Martini (1971) and Bukry (1978) argued for 1.8 Ma for the upper boundary of the *Discoaster brouweri* zone, while Bylinskaya and Golovina (2004) as well as Okada (2000) suggest 1.95 Ma, and Berggren and co-authors (1983a) and Browning et al. (2017) date it to 1.9 Ma.

Four *Discoaster* zones can be recognized in our core sections. The datum levels of four *Discoaster* species have been recently revised, notably by Bergen et al. (2019) and Browning et al. (2017), thus allowing the boundaries between corresponding zones of nominate taxa to be refined.

The *Discoaster brouweri* zone (2.29–1.9 Ma, in this study; or 2.39–1.93 Ma, by Backman et al. 2012) covers the interval between the LO of the species of this name and that of *Discoaster pentaradiatus.* The latter is estimated at 2.39 or 2.29 Ma by Backman et al. (2012) and Bergen et al. (2019), respectively. The FO of *Discoaster pentaradiatus* defines the lower boundary of the zone of this species (2.51–2.29 Ma, in this study; and 2.76–2.39 Ma, after Backman et al. 2012). In some cores, only a combined zone of *Discoaster brouweri*/*Discoaster pentaradiatus* can be documented (see Chap. 7).

The underlying *Discoaster surculus* zone (2.51–2.73 Ma, in this study, Fig. 2) in Gartner's scheme (1977) is not considered by some authors, for instance, by Backman et al. (2012). In our material, only a combined zone of *Discoaster surculus*/*Discoaster tamalis* was found in the lower part of three sediment cores (see Chap. 7). The LO of *Discoaster tamalis* is estimated at 2.73, 2.76 and 2.9 Ma by Bergen and co-authors (2019), Backman and co-authors (2012) and Rahman and Rothe (1989), respectively. The FO of the species is unknown. Browning et al. (2017) report the LO of the *Discoaster tamalis* Acme zone at 3.082 Ma, while Backman et al. (2012) place the lower boundary of the zone of this name at 3.82 Ma.

Unfortunately, the nannofossil preservation is generally not good enough to identify all the zones listed above; some are washed out from one or even all four of the core sections studied. This issue is discussed in Chap. 7. The most robust datum levels in our material appear to be the base of the *Emiliania huxleyi* zone at 0.27 Ma (or 0.29 Ma, according to Backman et al. 2012), the disappearance of discoasters at ~1.9 Ma (from 1.8 to 1.95 Ma, according to various authors) and the boundary of the *D. surculus*/*D. pentaradiatus* zones at 2.51 Ma (Bergen et al. 2019) or 2.53 Ma (Backman et al. 2012). The latter represents the nannofossil datum level in our material that is closest to the Pliocene/Pleistocene boundary (2.59 Ma, after Cohen et al. 2013, 2020).

4 Conclusions

Previously published stratigraphic schemes and datum levels of planktic foraminiferal and nannofossil index species from the Atlantic were compared to develop a stratigraphic framework for the Ioffe Drift area to cover the last 3.8 Ma. The ages of the zonal boundaries generally coincide with the index species' datum level, namely their first or last occurrence.

For the last 3.8 Ma, foraminiferal stratigraphy comprises eight zones and subzones; for convenience and consistency, in the text and figures they all are termed zones. The uppermost zone of *Globigerinoides ruber* (pink), previously defined by Barash and co-authors (1983) in DSDP Sites 516 and 518 from the Rio Grande Rise, is characteristic of the area, while the other zones are regional or even largely typical of the tropical–subtropical Atlantic.

For this same time span, the nannofossil stratigraphic framework consists of nine zones. The upper five have been adopted from the most detailed scheme by Gartner (1977), with later revisions to the age of some boundaries (e.g., Backman et al. 2012). The lower four zones are referred to by the name of the respective species of discoasters, taking into account the datum levels of the index species from recent publications (Browning et al. 2017; Bergen et al. 2019).

References

Backman J, Raffi I, Rio D et al (2012) Biozonation and biochronology of Miocene through Pleistocene calcareous nannofossils from low and middle latitudes. Newsletters Stratigr 45:221–244. https://doi.org/10.1127/0078-0421/2014/0042

Barash MS (1988) Quaternary paleooceanology of the Atlantic Ocean. Nauka, Moscow (in Russian)

Barash MS, Oskina NS, Blyum NS (1983) Quaternary biostratigraphy and surface paleotemperatures based on planktonic foraminifers. Initial reports DSDP, Leg 72, Santos, Brazil, pp 849–869. https://doi.org/10.2973/dsdp.proc.72.142.1983

Bergen JA, Truax S, de Kaenel E et al (2019) BP Gulf of Mexico Neogene astronomically tuned time scale (BP GNATTS). Bull Geol Soc Am 131:1871–1888. https://doi.org/10.1130/B35062.1

Berggren WA, Aubry MP, Hamilton N (1983a) Neogene magnetobiostratigraphy of deep-sea drilling project site 516 (Rio Grande Rise, South Atlantic). Initial reports DSDP, Leg 72, Santos, Brazil, pp 675–713. https://doi.org/10.2973/dsdp.proc.72.130.1983

Berggren WA, Hamilton N, Johnson DA et al (1983b) Magnetobiostratigraphy of deep-sea drilling project leg 72, sites 515–518, Rio Grande Rise (South Atlantic). In: Initial reports of the deep-sea drilling project, 72. U.S. Government Printing Office. https://doi.org/10.2973/dsdp.proc.72.149.1983

Berggren WA, Hilgen FJ, Langereis CJ et al (1995a) Late Neogene chronology: new perspectives in high-resolution stratigraphy. Geol. Soc. Amer. Bull. 107:1272–1287

Berggren WA, Kent DV, Swisher CC III, Aubry MP (1995b) A revised cenozoic geochronology and chronostratigraphy. In: Berggren WA, Kent DV, Hardenbol J (eds) Geochronology, time scales and global stratigraphic correlations: a unified temporal framework for an historical geology. Soc. Econ. Paleont. Min. special publication vol 54, pp 129–212. https://doi.org/10.1130/0016-7606(1985)96%3c1407:CG%3e2.0.CO;2

Blow WH (1969) Late Middle Eocene to Recent planktonic foraminiferal biostratigraphy. In: Brönnimann P, Renz HH (eds) Proceedings of the first international conference on planktonic microfossils, vol 1. E.J. Brill, Leiden, Geneva, pp 199–422

Bolli HM, Premoli Silva I (1973) Oligocene to recent planktonic foraminifera and stratigraphy of the Leg 15 Sites in the Caribbean Sea. In: Initial reports of the deep-sea drilling project, 15. U.S. Government Printing Office

Bolli HM, Saunders JB (1985) Oligocene to Holocene low latitude planktonic foraminifera. In: Plankton stratigraphy. Cambridge University Press, pp 155–262

Browning E, Bergen J, Blair S et al (2017) Late Miocene to Late Pliocene taxonomy and stratigraphy of the genus Discoaster in the circum North Atlantic Basin: Gulf of Mexico and ODP Leg 154. J Nannoplankton Res 37(2–3):189–214

Bukry D (1978) Biostratigraphy of Cenozoic marine sediments by calcareous nannofossils. Micropaleontology 24:44–60. https://doi.org/10.2307/1485419

Bylinskaya ME, Golovina LA (2004) Correlation of the Pliocene–Quaternary foraminiferal and nannofossil zonations in the North Atlantic. Stratigr Geol Correl (in Russian with English trans) 12:309–319

Chaisson WP, Pearson PN (1997) Planktonic foraminifer biostratigraphy at Site 925: middle Miocene–Pleistocene. In: Proceedings of the ocean drilling program, 154 scientific results. Ocean Drilling Program

Cohen KM, Finney SC, Gibbard PL, Fan J-X (2013) The ICS international chronostratigraphic chart. Episodes 36. https://doi.org/10.18814/epiiugs/2013/v36i3/002

Cohen KM, Finney SC, Gibbard PL, Fan J-X (2020) The ICS international chronostratigraphic chart. Episodes 36. (update). http://www.stratigraphy.org/ICSchart/ChronostratChart2020-01.pdf

Dmitrenko OB (1987) A detailed zonal scale of the Quaternary bottom deposits based on coccoliths (on the Rio Grande Rise in the Atlantic Ocean). Oceanol (in Russian with English trans), 27:460–464

Gartner S (1977) Calcareous nannofossil stratigraphy and revised zonation of the Pleistocene. Mar Micropal 2:1–25

Ivanova E, Murdmaa I, Borisov D, Dmitrenko O, Levchenko O, Emelyanov E (2016) Late Pliocene-Pleistocene stratigraphy and history of formation of the Ioffe calcareous contourite drift, western South Atlantic. Mar Geol 372:17–30. https://doi.org/10.1016/j.margeo.2015.12.002

Ivanova E, Borisov D, Dmitrenko O, Murdmaa I (2020) Hiatuses in the late Pliocene–Pleistocene stratigraphy of the Ioffe calcareous contourite drift, western South Atlantic. Mar Pet Geol 111. https://doi.org/10.1016/j.marpetgeo.2019.08.031

Kennett JP, Srinivasan MS (1983) Neogene planktonic foraminifera. A phylogenetic atlas. Hutchinson Ross, Stroudsburg, PA

Leonard KA, Williams DF, Thunell RC (1983) Pliocene paleoclimatic and paleoceanographic history of the South Atlantic Ocean: stable isotopic records from Leg 72 deep-sea drilling project holes 516A and 517. In: Barker PF, Carlson RL, Johnson DA et al (eds) Initial reports of DSDP, Leg 72. Government Printing Office, Washington, U.S, pp 895–906

Lourens L, Hilgen F, Shackleton NJ et al (2004) The Neogene period. In: Gradstein F, Ogg J, Smith A (eds) A geologic time scale 2004. Cambridge University Press Cambridge, UK, pp 409–440

Martini E (1971) Standard tertiary and quaternary calcareous nannoplankton zonation. In: Proceedings of the second international conference on planktonic microfossils Rome 2

Okada H (2000) Neogene and quaternary calcareous nannofossils from the Blake Ridge, Sites 994, 995, and 997. In: Proceedings of the ocean drilling program, 164 scientific results. Ocean Drilling Program

Okada H, Bukry D (1980) Supplementary modification and introduction of code numbers to the low latitude coccolithbiostratigraphic zonation (Bukry, 1973, 1975). Mar Micropaleontol 5. https://doi.org/10.1016/0377-8398(80)90016-X

Rahman A, Roth PH (1989) Late Neogene calcareous nannofossil biostratigraphy of the Gulf of Aden region based on calcareous nannofossils. Mar Micropaleontol 15:1–27

Thierstein HR, Geitzenauer KR, Molfino B, Shackleton NJ (1977) Global synchroneity of late quaternary coccolith datum levels: validation by oxygen isotopes. Geology 5:400–404. https://doi.org/10.1130/0091-7613(1977)5%3c400:GSOLQC%3e2.0.CO;2

Wade BS, Pearson PN, Berggren WA, Pälike H (2011) Review and revision of Cenozoic tropical planktonic foraminiferal biostratigraphy and calibration to the geomagnetic polarity and astronomical time scale. Earth-Science Rev 104:111–142. https://doi.org/10.1016/j.earscirev.2010.09.003

General Methods

Dmitrii Borisov and Elena Ivanova

Abstract The chapter describes the main datasets (and methods for collecting them) that were used for the comprehensive study of the Ioffe Drift.

1 Sub-bottom profiling and seismic data

A very high-resolution sub-bottom profiling (center frequency 4–7 kHz) was carried out during the R/V *Akademik Ioffe* Cruises 32 and 52 (2010 and 2016) using the parametric echo-sounder *SES 2000 deep* (Levchenko & Murdmaa 2013; Ivanova et al. 2018). The acquired profiles run across the summit and northern slope of the Ioffe Drift and the series of wave-like depositional features to the WSW from the drift (Fig. 1a, b). Sound velocity, both in water and in the upper part of the sediment section, was taken as 1,500 m/s. Access to the seismic records collected from the study area during the R/V *Vema* Cruise 18, Leg 4 (1962) (Fig. 1b) was provided to the authors via the Marine Geoscience Data System (GeoMapApp, www.geomapapp.org/CC BY).

2 Bathymetry

Bathymetric resources for the study included gridded bathymetric data from GEBCO 2020 for the southern Brazil Basin (GEBCO Compilation Group 2020), a physiographic map of the Brazil oceanic margin (scale 1:1,883,927) prepared by the Navy Center of Hydrography of the Brazil Directorate of Hydrography and Navigation (*Diretoria de Hidrografia e Navegação, Marinha do Brasil* 2017), and the results

D. Borisov (✉) · E. Ivanova
Shirshov Institute of Oceanology, Russian Academy of Sciences, Moscow, Russia
e-mail: dborisov@ocean.ru

E. Ivanova
e-mail: e_v_ivanova@ocean.ru

I. Murdmaa and E. Ivanova (eds.), *The Ioffe Drift*, Springer Geology,
https://doi.org/10.1007/978-3-030-82871-4_4

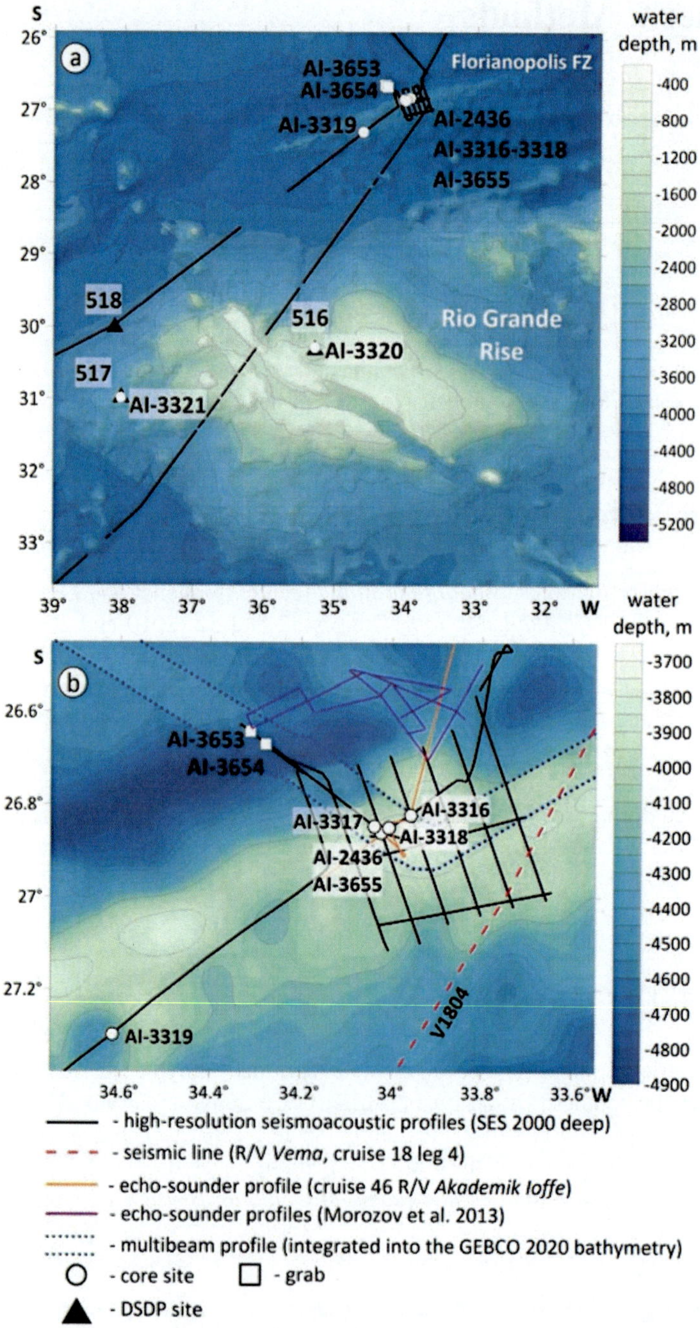

S
a
Florianopolis FZ
AI-3653
AI-3654
AI-3319
AI-2436
AI-3316-3318
AI-3655
518
516
AI-3320
Rio Grande Rise
517
AI-3321
26°
27°
28°
29°
30°
31°
32°
33°
39°
38°
37°
36°
35°
34°
33°
32°
W
water depth, m
-400
-800
-1200
-1600
-2000
-2400
-2800
-3200
-3600
-4000
-4400
-4800
-5200
b
S
26.6°
26.8°
27°
27.2°
AI-3653
AI-3654
AI-3317
AI-3316
AI-3318
AI-2436
AI-3655
AI-3319
V1804
34.6°
34.4°
34.2°
34°
33.8°
33.6°W
water depth, m
-3700
-3800
-3900
-4000
-4100
-4200
-4300
-4400
-4500
-4600
-4700
-4800
-4900
- high-resolution seismoacoustic profiles (SES 2000 deep)
- seismic line (R/V Vema, cruise 18 leg 4)
- echo-sounder profile (cruise 46 R/V Akademik Ioffe)
- echo-sounder profiles (Morozov et al. 2013)
- multibeam profile (integrated into the GEBCO 2020 bathymetry)
- core site
- grab
- DSDP site

◀**Fig. 1** **a** Bathymetric map of the western segment of the Rio Grande Rise and the ridge within the Florianopolis Fracture Zone, showing the locations of the core sites and high-resolution seismoacoustic (sub-bottom) profiles; **b** Detailed bathymetric map of the Ioffe Drift showing the locations of the core and grab sites, high-resolution seismoacoustic (sub-bottom) profiles, single-beam echo-sounder profiles, multibeam profile (GEBCO Compilation Group 2020) and seismic profile acquired during the R/V *Vema* cruise 18, Leg 4

of the high-resolution sub-bottom profiling mentioned above. In addition, hydrographic surveys were carried out during Cruise 46 of the R/V *Akademik Ioffe* (2014) as well as Cruises 36 and 37 of the R/V *Akademik Sergey Vavilov* (2012, 2013) using a single-beam echo-sounder ELAC LAZ 4700 (12 kHz) with a STG 721 digitizer (Morozov et al. 2013; Ivanova et al., 2016a). The locations of the very high-resolution sub-bottom profiles and single-beam echo-sounding profiles are shown in Fig. 1b.

3 Sediment Cores

The site AI-2436 at the drift summit was chosen on the basis of the sub-bottom profiling data, and this first core was collected during the R/V *Akademik Ioffe* Cruise 32 using a 6 m-long gravity corer with an outer/inner diameter of 127/110 mm (Table 1; Ivanova et al. 2016b). As the corer was completely filled with sediment, the actual core was initially 6 m long; however, as we had to apply hydraulic pressure to push the core out, it was artificially lengthened to 7.14 m, with corresponding thinning of the core diameter in its upper parts. As this increase could not be accurately estimated through the various parts of the core, in the following description, sampling and study we regard it as measuring 7.14 m.

The gravity cores AI-3316 to AI-3319 were retrieved during Cruise 46 of the same vessel in 2014 in the aforementioned profiles (Fig. 1), using an 8 m-long gravity corer with an outer/inner diameter of 127/110 mm. Core AI-3318 was taken from the drift summit to obtain a more appropriate record of the uppermost sediment cover than the core AI-2436, whose upper part was stretched as described (Ivanova et al. 2016a, 2020). Later, during the R/V *Akademik Ioffe* Cruise 52 in 2016, another core (AI-3655) was collected from the same location to refine both the stratigraphy and other results (Ivanova et al. 2018). During this latter cruise, two grabs, AI-3653 and AI-3654, were retrieved from the deep fault-controlled valley at the base of the drift's northern slope to investigate its floor (Fig. 1). Cores AI-3316 and AI-3317 were taken from the northern slope of the drift on the Antarctic Bottom Water and Lower Circumpolar Deep Water pathway to the east along the ridge covered by drift (Fig. 1b). Thus, the core locations permit a comparison of the processes acting on the summit and on the upper northern slope. In addition, core AI-3319 was collected to the west-southwest from the Ioffe Drift. During the R/V *Akademik Ioffe* Cruise 46, the cores AI-3320 and AI-3321 were collected from the summit and slope of the

Table 1 Coordinates, water depth, core length and physiographic environment of sediment cores from the Ioffe Drift and Rio Grande Rise. For locations, see Figure 1a, b

Core	Latitude	Longitude	Depth (m)	Length (m)	Location
AI-2436*	26°51.60′S	34°01.40′W	3799	7.14**	Summit of the Ioffe drift
AI-3316***	26°49.33′S	33°57.46′W	3898	4.89	Northern slope of the Ioffe drift
AI-3317	26°50.79′S	34°02.33′W	3832	5.15	Northern slope of the Ioffe drift
AI-3318***	26°50.98′S	34°00.41′W	3788	3.46	Summit of the Ioffe drift
AI-3319	27°17.79′S	34°37.08′W	4066	2.87	Series of wave-like depositional features to WSW from the Ioffe drift
AI-3320	30°16.50′S	35°17.12′W	1323	1.16	Summit of the Rio Grande Rise (at DSDP Site 516)
AI-3321	30°56.85′S	38°02.45′W	2969	2.93	Slope of the Rio Grande Rise (at DSDP Site 517)
AI-3655	26°51.59′S	34°01.40′W	3799	5.14	Summit of the Ioffe drift

Notes *From Ivanova et al. (2016b), corrected. **The upper part of the core was stretched during processing. ***From Ivanova et al. (2020)

Rio Grande Rise at DSDP Sites 516 and 517, respectively (Table 1; Fig. 1a; Ivanova et al. 2016a).

All cores were opened, described and continuously sampled onboard. During and after the cruises, the 1–2 cm thick samples (in various cores) were kept in a storage room at 4 °C.

In cores AI-3316 to AI-3318 the relative abundance of several components, notably nannofossils, and the types of sediment were examined in smear-slides from throughout the sections under the onboard polarizing microscope.

Cores AI-2436, AI-3318 and AI-3316 have been investigated previously as a significant part of the scientific results published in Ivanova et al. (2016b, 2020); however, as well as presenting new data, this monograph serves as a partial revision of these publications (see Chaps. 6, 7, 8 and 9). The scientific results obtained from sediment cores AI-3655, AI-3317 and AI-3319 (Table 1) are presented here for the first time. All results pertaining to grain size and mineralogy are new also those regarding the ferromanganese nodules (see Chaps. 6 and 8).

References

Diretoria de Hidrografia e Navegação, Marinha do Brasil (2017) Fisiografia da margem brasileira – Modelo Digital do Terreno

GEBCO Compilation Group (2020) GEBCO 2020 grid

Ivanova E, Borisov D, Dmitrenko O, Murdmaa I (2020) Hiatuses in the late Pliocene-Pleistocene stratigraphy of the Ioffe calcareous contourite drift, western South Atlantic. Mar Pet Geol 111:624–637. https://doi.org/10.1016/j.marpetgeo.2019.08.031

Ivanova E, Murdmaa I, Borisov D et al (2016) Investigation of contourite systems in the South Atlantic during Cruise 46 of the R/V Akademik Ioffe. Oceanology 56:322–324. https://doi.org/10.1134/S0001437016040044

Ivanova E, Murdmaa I, Borisov D et al (2016) Late Pliocene-Pleistocene stratigraphy and history of formation of the Ioffe calcareous contourite drift, Western South Atlantic. Mar Geol 372:17–30. https://doi.org/10.1016/j.margeo.2015.12.002

Ivanova EV, Murdmaa IO, Borisov DG et al (2018) Geological and geophysical investigation of contourite systems from the Central and Southern Atlantic during Cruise 52 of the R/V Akademik Ioffe. Oceanology 58:337–339. https://doi.org/10.1134/S0001437018020054

Levchenko OV, Murdmaa IO (2013) Strategy of the lithological and seismoacoustic research of the deep-water deposits along transatlantic geotraverses during Cruise 32 of the R/V Akademik Ioffe in the autumn of 2010 (Kaliningrad to Ushuaia). Oceanology 53:124–128. https://doi.org/10.1134/S0001437013010098

Morozov EG, Demidova TA, Grigorenko KS et al (2013) Measurements of bottom currents in underwater channels of the Atlantic during Cruise 36 of the R/V Akademik Sergey Vavilov. Oceanology 53:762–764. https://doi.org/10.1134/S0001437013060064

The Geomorphology and Seismic Structure

Dmitrii Borisov, Oleg Levchenko, and Natalia Libina

Abstract The study of geomorphology and seismic structure of the Ioffe Drift area was undertaken on the basis of high-resolution sub-bottom profiling data, an old low-resolution single-channel seismic profile, and single-beam and multibeam bathymetric data. The drift occupies the central part of a tectonic ridge related to the Florianopolis (Rio Grande) Fracture Zone. The drift's morphology and orientation, which are significantly controled by the basement relief, were also strongly affected by bottom currents. The acoustic penetration of 60–80 m allowed detailed investigation of only the upper part of the drift deposits. Four seismic units were identified in the acoustic structure of the drift. Changes in the seismic characteristics of the units represent changes in bottom conditions during the drift's formation from tranquil to highly dynamic. Evidence of drastic erosion permeates the entire structure of the drift. The major unconformities separating the units correspond to the most prominent erosional events in the drift's history. The absence of data does not allow the age of these unconformities to be identified. The correlation between seismic records collected from the drift and from the Rio Grande Rise demonstrates the lower impact of bottom currents on sedimentation across the rise.

1 Bathymetry

The geomorphologic study of the Ioffe Drift is based on the GEBCO global elevation model (GEBCO Compilation Group 2020), in conjunction with scarce multibeam data, single-beam echo-sounder profiles and high-resolution sub-bottom profiles. For

D. Borisov (✉) · O. Levchenko · N. Libina
Shirshov Institute of Oceanology, Russian Academy of Sciences, Moscow, Russia
e-mail: dborisov@ocean.ru

O. Levchenko
e-mail: olevses@mail.ru

N. Libina
e-mail: lnatvit@mail.ru

I. Murdmaa and E. Ivanova (eds.), *The Ioffe Drift*, Springer Geology,
https://doi.org/10.1007/978-3-030-82871-4_5

a detailed description of the available bathymetric data and location of the profiles, see Chap. 4.

The Ioffe Drift covering the ridge within the Florianopolis Fracture Zone is traced at a water depth ranging from 3,750 to 4,170 m. Its overall geometry changes from mounded and narrow in the western part of the study area to gentler, wider and smoother in the east (Figs. 1a and 2). A clear cross-sectional asymmetry characterizes the morphology of the drift. The northern slope of the ridge below 4,050 m water depth and beyond the northern limit of the drift shows an angle of 5.3°.

The analysis of multibeam data revealed an almost horizontal terrace 8.5 km wide at the base of the northern ridge slope, separating the ridge from the (channel-like) fault-controlled valley (Fig. 1b, c). The terrace is traced at a water depth ranging from 4,450–4,500 m. In the east, the terrace terminates at approximately 34°W. The elongation of the terrace toward the west cannot be evaluated due to insufficient data. The terrace has an irregular surface with several distinctive mounds 25–40 m in height, oriented parallel to the deep valley. The distance between mound crests is 2.5 km. The asymmetric fault valley, with a relative depth of 300 m, has a slope angle of up to 5° on the northern flank and 0.8° on the southern (Fig. 1c). The valley width exceeds 55 km (at the prominent slope-break near 4,500 m). The gentle valley flank to the left, in the direction of the Antarctic Bottom Water (AABW) flow, the steeper flank and the adjacent terrace to the right have a configuration similar to the Vema Channel and its eastern terrace, revealed at depths of 4,100–4,300 m (Johnson 1984). The eastern terrace is considered to have been formed under the influence of bottom currents, yet the particular mechanism of the formation remains unknown (Johnson 1984).

To the southwest of the drift (near core site AI-3319), the base of the ridge covered by the drift is marked by a U-shaped valley with a relative depth of 120 m and a width of approximately 15 km along the profile (Fig. 1d). The bottom of the valley was recorded at a water depth of 4,140 m.

2 Acoustic Structure of the Drift

The deep seismic structure of the Ioffe Drift remains understudied due to a lack of high-quality seismic data. The seismic records collected during the R/V *Vema* Cruise 18, Leg 4 (1962) revealed a large, mounded depositional body covering the basement of the ridge studied within the Florianopolis Fracture Zone (see Figs. 1b and 5.2). Its thickness varies from 0.3 s TWT (two-way travel time) over the ridge summit to 0.54 s TWT over its southern slope. The body does not cover the steeper northern ridge slope, where older deposits (or basement units) outcrop. The basement irregularities are only slightly pronounced in the overall geometry of the sediment body (Fig. 2). The low resolution and quality of these seismic data do not permit a detailed investigation of the sediment body structure.

The upper part of the drift was investigated using the *SES-2000 deep* parametric profiler. This provided acoustic penetration of 60–80 mbsf (meters below sea floor)

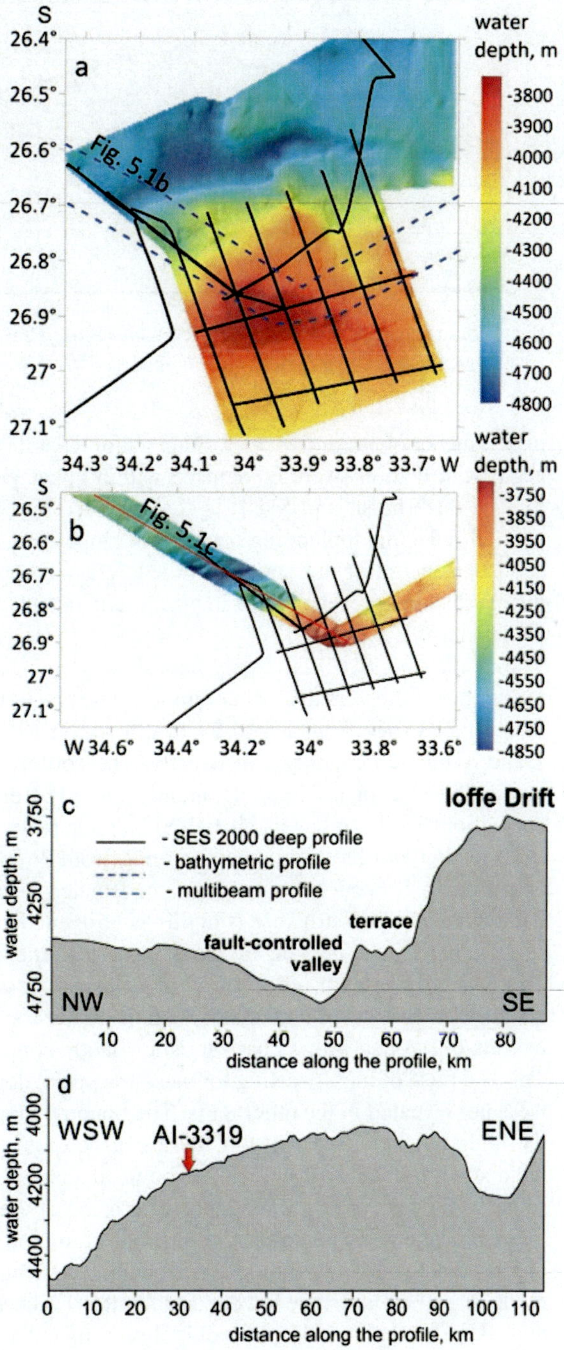

Fig. 1 **a** Detailed bathymetric map of the study area with location of sub-bottom (SES 2000 deep) and multibeam profiles; **b** Multibeam bathymetry (GEBCO Compilation Group 2020); **c** Bathymetric profile based on multibeam data; **d** Bathymetric profile (based on SES 2000 deep data) crossing the area to the WSW from the drift. Location of core AI-3319 is shown in Fig. 1

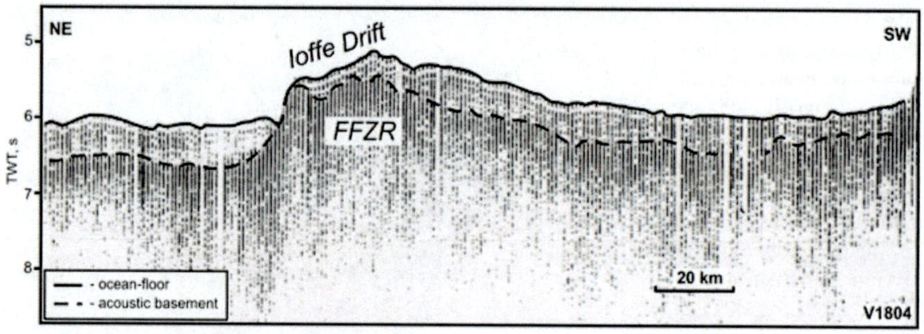

Fig. 2 Seismic profile collected during RV *Vema* (1962), Cruise 18, Leg 4, with interpretation. The location of the profile is shown in Fig. 1b

in sediments of the Ioffe Drift. Four major seismic units, demonstrating different acoustic structures, were recognized within this depth interval (Figs. 1, 2, 3, 4 and 5). The lowermost, unit SU-1, is characterized by moderate-amplitude continuous parallel reflectors toplapping on the drift slopes (Figs. 1, 2, 3, 4 and 5). Insufficient acoustic penetration does not allow the lower unit boundary to be determined, yet its visible thickness reaches 35 m in places and up to 12 reflectors can be traced within the layer.

Seismic unit SU-2 is mostly acoustically transparent. There are several weak intermittent reflectors that occur only in some places within this layer. The thickness of the unit varies from 2.5 to 27 m (14 m, on average). Both seismic units SU-1 and SU-2 lie generally consistently. The conformal boundary between them is marked by a strong reflector, apparently caused by erosion (in some places) or non-deposition. By contrast, seismic unit SU-3 is characterized by a well-defined acoustic stratification and has a thickness varying from 36 m to complete pinching (16 m, on average). It occurs in some areas as isolated lens-shaped subunits (Fig. 4a, b). Its layered acoustic structure is highly heterogeneous compared to that of the other seismic units. It is created by closely spaced reflectors numbering from 4 to 14. The generally smooth reflectors in places have an undulating character (Figs. 3b, c and 5c). The structure of the SU-3 unit on the southern drift slope is interrupted by lens-like zones with a characteristic chaotic configuration of internal reflection. This is typical of mass-wasting (or mass-transport) deposits. Similar seismic patterns were not revealed in the other units. The boundary between stratified unit SU-3 and the underlying transparent unit SU-2 can be quite confidently determined where the distinct reflectors of unit SU-3 show an apparent onlap on top of unit SU-2. In some seismic records, the evidence of erosion at the boundary between SU-2 and SU-3 is unclear (Figs. 3a, b and 4a). Furthermore, a smooth transition zone was observed between these units, corresponding to a smooth change in sedimentation mode. In some places, the lower part of unit SU-3 becomes similar to the underlying unit SU-2, so the boundary is barely discernible. The upper boundary of unit SU-3 is confidently determined by a distinct and expressive angular unconformity that

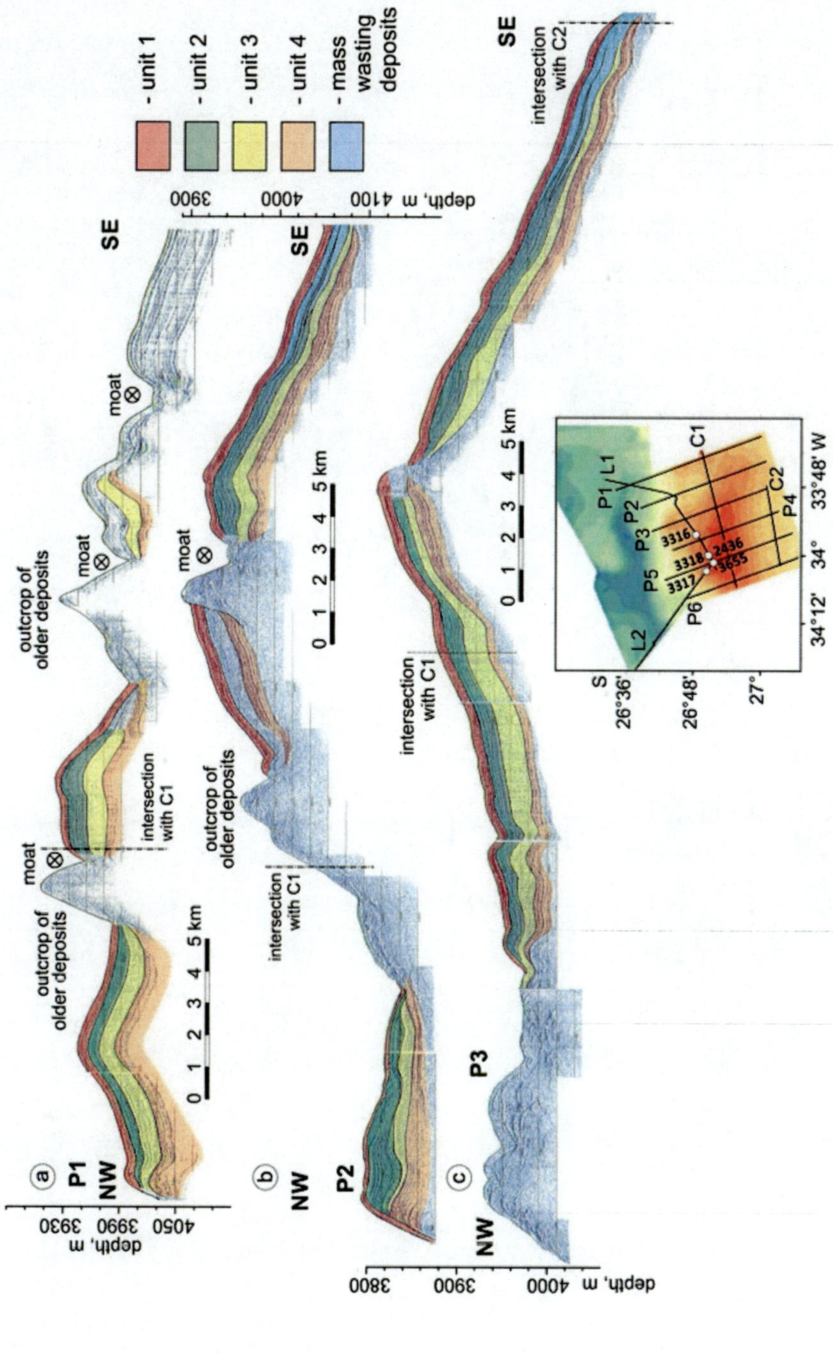

Fig. 3 High-resolution sub-bottom profiling records (SES-2000 deep), with interpretation. The cross in a circle marks the eastward direction of the bottom current. The locations of the profiles are shown in the inset and Fig. 1

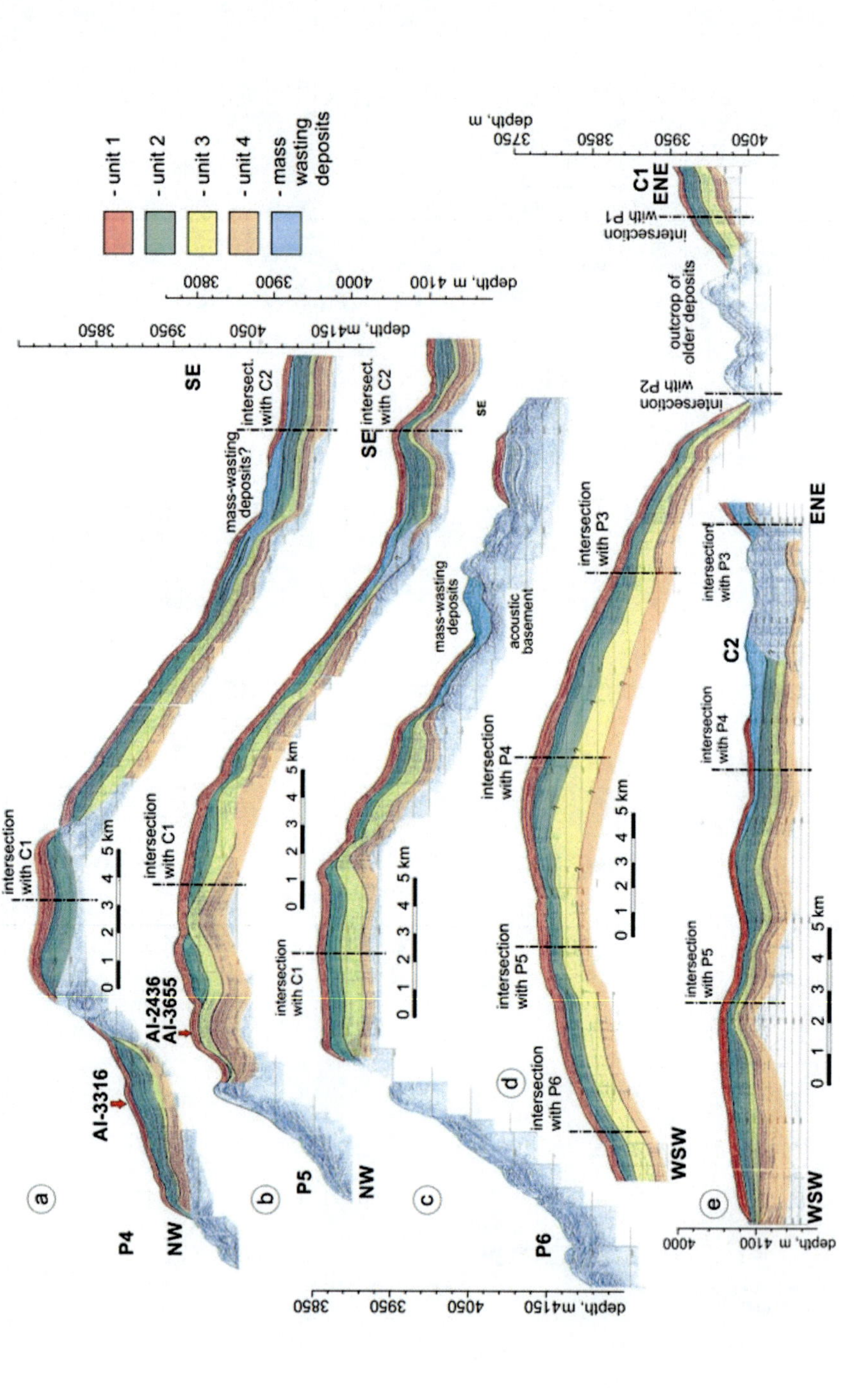

Fig. 4 High-resolution sub-bottom profiles (with interpretation) collected from the Ioffe Drift using the SES-2000 deep profiler. Arrows mark location of core sites. Location of the profiles is shown in Fig. 1

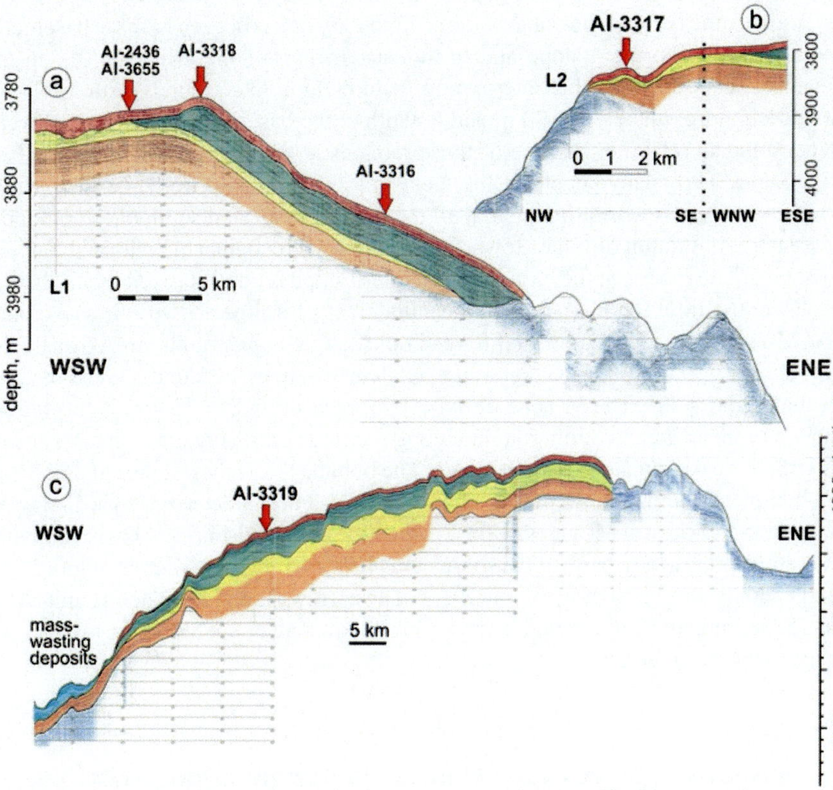

Fig. 5 High-resolution sub-bottom profiles (with interpretation) collected from: **a** and **b** Ioffe Drift; and **c** base of the southern slope of the drift-covered ridge. Arrows mark the location of core sites. Location of the profiles is shown in Fig. 1

truncates the reflectors. This unconformity also represents the upper boundary of unit SU-2 (Figs. 4b and 5a, b) in the areas where unit SU-3 wedges out.

The uppermost unit, SU-4, is characterized by high- to moderate-amplitude continuous parallel reflectors conformal to the surface of the seafloor. In addition, several zones of intermittent reflectors were revealed in various parts of the drift (Fig. 5a). The number of reflectors recognized varies from 3 to 9. The unit thickness varies between 4 and 21 m (10 m, on average).

The same acoustic structure was recorded in the area near core site AI-3319 (southwest of the Ioffe Drift). This area is characterized by a decreased thickness of unit SU-4 not exceeding 10 m. The thickness of units SU-3 and SU-2 varies from 5 to 38 m. The reflectors within all the units have an irregular, wavy character. These undulations are clearly expressed in the bottom surface. The height of the resulting wave-like depositional features does not exceed 10 m. To the south-west of the core site AI-3319, the acoustic structure of the deposits has been strongly disturbed by mass-wasting processes (Fig. 5c).

The seismic records show that some sedimentary units (or even basaltic basement) outcrop on the northern slope and in the eastern part of the drift (Fig. 2). In the seismoacoustic records, the outcropping features look like mounds with a relative height of approximately 50–90 m and a width at the base of 2–4 km. Several well-defined moats relate to the base of these mounds at depths of 3,940–4,030 m (see Fig. 3a, b). The relative depth of the moats varies from 5–40 m, while their width reaches 2 km. It is worth to note that all the moats are located south of the mounds. The lateral migration of moats toward the mounds is recognizable within the upper two units. The terrace at the base of the southern slope is characterized by its chaotic acoustic structure. Detailed study of the units' thickness allowed two zones (A-zones and M-zones) to be distinguished for each of the three upper units. In A-zones, the unit thickness exceeds the average value, while in M-zones the unit thickness reaches its maximum value. The M-zone of unit SU-4, measuring 8 × 15 km, is located on the top of the drift, extending roughly along the ridge crest (Fig. 6a). The A-zone of this unit is also elongated along the ridge. The boundaries of these zones of unit SU-3 are completely unlike those of unit SU-4. Unit SU-3 is characterized by a U-shaped A-zone and three oval M-zones (12 × 7.5 km, 4 × 8 km and 14.5 × 4 km) occupying the central and eastern parts of the survey area (Fig. 6b). The boundaries of unit SU-2 are generally similar to those of unit SU-4. The narrow A- and M-zones of unit SU-2 (4–15 km long and 2–6 km wide) extend for 38 km and 27 km along the ridge crest, respectively (Fig. 6c).

3 Seismic Evidence of the Contourite Origin of the Drift

A number of seismic criteria was developed for contourite recognition (Faugères et al. 1999; Rebesco and Stow 2001; Nielsen et al. 2008; Rebesco et al. 2014). One of these criteria is an upwardly convex geometry that is nonparallel to the accumulation surface created by the preceding erosional event (Faugères et al. 1999). Unfortunately, this surface lies deeper than the maximum penetration depth achievable by the *SES-2000 deep* profiles (80 m). Neither does the low resolution of the seismic record collected during the R/V *Vema* Cruiser 18, Leg 4 (1962) permit this surface to be distinguished. Analysis of the acquired seismoacoustic data revealed further evidence of the impact of a strong bottom current on the drift's formation. According to the morphostructural context, general morphology and local hydrodynamic setting, the depositional body under investigation can be classified as an elongated mounded drift. Although it meets the large-scale criterion of down-current elongation, its overall geometry and orientation seem to be mainly predefined by the relief of the underlying basalt basement.

The upper part of the drift sediment cover is characterized by broadly lenticular, upwardly convex seismic units with downlapping, onlapping and toplapping reflectors, considered to be medium-scale diagnostic criteria for contourites (Rebesco and Stow 2001). The boundaries between the seismic units that were distinguished

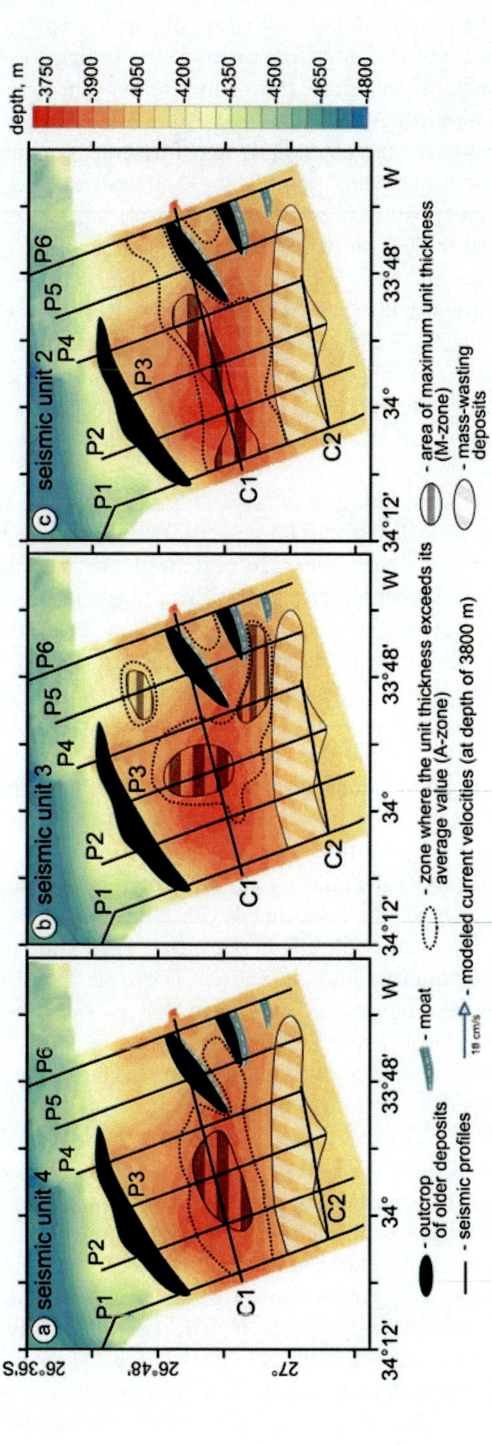

Fig. 6 **a–c** Distribution of the thickness of seismic units SU-4, SU-3 and SU-2, respectively, including areas of maximum unit thickness and areas where the unit thickness exceeds the average value

meet the definition of major discontinuities within the drift, a criterion of a significant impact by bottom currents on drift formation. According to Faugères et al. (1999), these discontinuities can represent timelines in the drift's evolution history and correspond to hydrological events.

The relative positions of the mounded outcrops of basement or ancient sedimentary rocks, the moats (zones of erosion) south of these mounds and the adjacent zones of deposition south of the moats suggest that, in terms of time, space and direction, the long-term activity of the bottom-water flow was quite stable (Fig. 3a, b). This configuration of erosional and depositional zones at the base of topographic obstacles is described in many articles (Faugères et al. 1999; Rebesco and Stow 2001; Hernández-Molina et al. 2008; Rebesco et al. 2014) and is considered to be evidence of the influence of bottom currents on sedimentation. In the present geomorphological and oceanographic context, the lateral migration of erosion and deposition zones (influenced by the Coriolis force) meets the criterion stated by Rebesco and Stow (2001).

The outcrops of older deposits (or basement) appear to play an important role in moat formation on the Ioffe Drift. The velocity of the flow increases around such topographic obstacles and the Coriolis force deflects the water flow toward the outcrop (to the left, in the southern hemisphere). This process results in erosion and the formation of moats (referred to as erosional valleys in some articles) close to the mounded outcrops. The orientations and relative positions of the mounds and moats imply an easterly direction of the bottom current, and the water depth interval and suggested direction correspond to those of the Lower Circumpolar Deep Water (LCDW) current. The maps of unit thickness and the acoustic structure of three upper seismic units both imply changes to the mode of sedimentation during the drift's formation.

A comparison between the seismic data and the results of numerical simulations using the INMOM model (Dianskii et al. 2002) demonstrate that the zone of maximum thickness of unit SU-4 is associated with the center of the gyre formed by the LCDW current (Ivanova et al. 2020). This zone marks the location of the depocenter during the formation of this unit. The absence of clear evidence of erosion within its seismoacoustic structure does not mean that erosional processes were not significant. The uppermost part of the drift's sediment record is interrupted by several long-term hiatuses, described in detail in Chap. 9. These hiatuses are generally associated with periods of bottom-current intensification (Ivanova et al. 2016, 2020).

Of all the units identified, unit SU-3 has both the greatest thickness and the clearest evidence of deep erosion. Slope instability resulted in the formation of mass-wasting deposits within its structure (Figs. 3b, c, 4a–c, e and 5c) and might have been caused by high sedimentation rates and tectonic movements (the study area is located within the fracture zone) acting as a trigger for mass-wasting processes. Sediment composition could also affect slope stability. The Ioffe drift deposits mainly composed of nannofossils and foraminifera tests are characterized by low shear-stress values and small slope angles sufficient to initiate gravity-driven sedimentation processes. The analysis of the seismic data revealed at least two mass-wasting events in the seismic sections, appearing as lens-like intrusions into unit SU-3. The specific configuration of the zone of maximum thickness is probably the result of intense erosion. The unconformity separating units SU-3 and SU-4 is considered to be due to the deepest erosion observed in the seismic sections; in several zones, unit SU-3 is totally eroded (Fig. 5a, b).

The acoustic structure of unit SU-2 reflects more tranquil conditions. Low-amplitude intermittent reflectors, acoustically transparent zones and the extension of the area of maximum unit thickness along the drift and in the main direction of the Lower Circumpolar Water (LCPW) imply that the sedimentation took place under the influence of weak bottom currents, with periods when pelagic settling prevailed. The elongated A- and M-zones and the relatively simple, parallel-layered structure of unit SU-2 point to its accumulation under the stable conditions of a downstream current. The shapes of the unit's A- and M-zones suggest that there was little or no impact by the LCPW gyre on sedimentation during the deposition of this unit (unlike unit SU-3). The well-stratified structure of unit SU-1 seems to be similar to that of units SU-3 and SU-4 and probably corresponds to a period when sedimentation was generally controled by bottom currents.

The acoustic structure of upper sediment cover recorded down to 60–80 msbf in the *SES-2000 deep* profiles of the Ioffe Drift differ significantly from those on the Rio Grande Rise and near the Vema Channel. A fragment of the *SES-2000 deep* profile running through the Vema Channel and the Vema Terrace is shown in Fig. 7. The acoustic penetration in the channel is approximately 27 m (Fig. 7a). The upper sedimentary layer, 15 m thick, is acoustically transparent, with two strong reflectors at depths of 7 mbsf and 11mbsf. These appear to be dense, coarse-grained sediments that were eroded and redeposited by the AABW current. Fine-grained

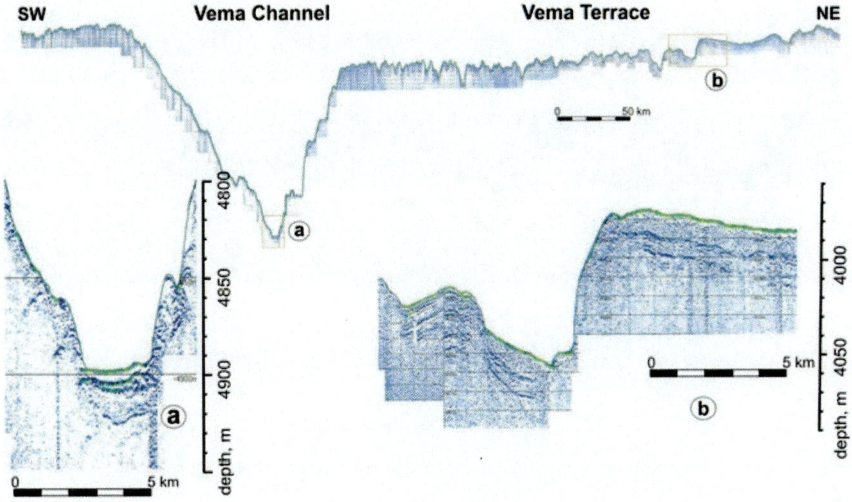

Fig. 7 High-resolution sub-bottom profiles (SES-2000 deep) running through the Vema Channel and Vema Terrace

unconsolidated sediments were swept by this current from the channel to the north. The chaotic acoustic structure of the lower sediments is similar to that of deposits on the flanks of the channel. The rugged topography of the flanks seems to be of an erosive nature, caused by this current's activity.

The acoustic penetration on the eastern Vema Terrace reaches 60 m. The sharp relief, with the amplitude of up to 100 m, is due to small, alternating ridges and troughs several kilometers across (Fig. 7), unlike the smooth topography seen in the low-resolution single- and multichannel seismic profiles in this region (Gamboa et al. 1983; Johnson 1984). The acoustic structure is characterized by stratified deposits of variable thickness (20–60 m) and uncommon acoustically transparent or chaotic deposits.

The Vema Terrace and the Vema Channel (eastern branch) extend around the northern flank of the rise, where the Florianopolis Fracture Zone controls their position. The easterly flow of AABW and LCDW through the Florianopolis Fracture Zone can interact with gravity flows transporting sediment material from the Rio Grande Rise to the Brazil Basin.

The *SES-2000 deep* profiles show the change in the acoustic structure of the upper sediment cover upon the transition from the fault-controlled valley of the Florianopolis Fracture Zone to the Ioffe Drift (Figs. 8 and 9). The profiles show no sub-bottom penetration in the sediments of the valley. The deposits within the fracture zone are characterized by a chaotic and opaque acoustic structure unlike that of the stratified deposits of the Ioffe Drift. This difference seems to be caused by the lithology of the sediments and the bottom's topography, which is rough in the fault zone and smooth in the drift. The sediment cover on the Rio Grande Rise slopes is represented mainly by opaque or chaotic deposits of the down-slope density turbidity flows and

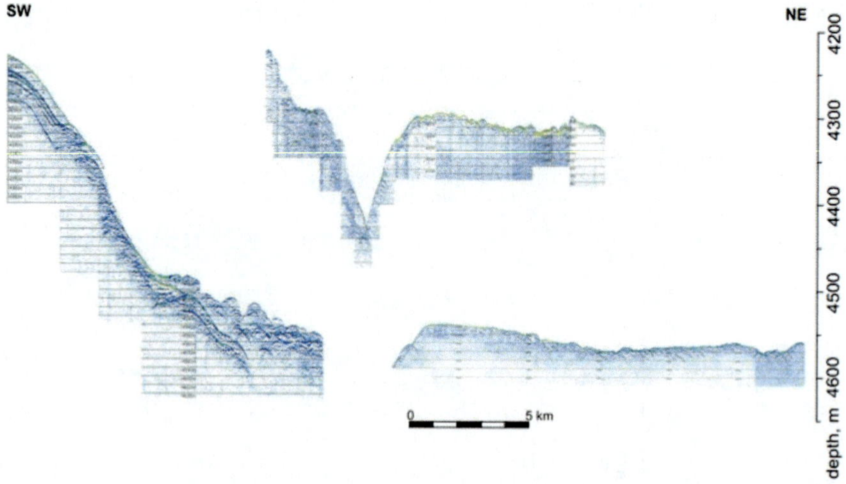

Fig. 8 High-resolution sub-bottom profiles (SES-2000 deep) running from the fault-controlled valley of the Florianopolis Fracture Zone to the Ioffe Drift to the north

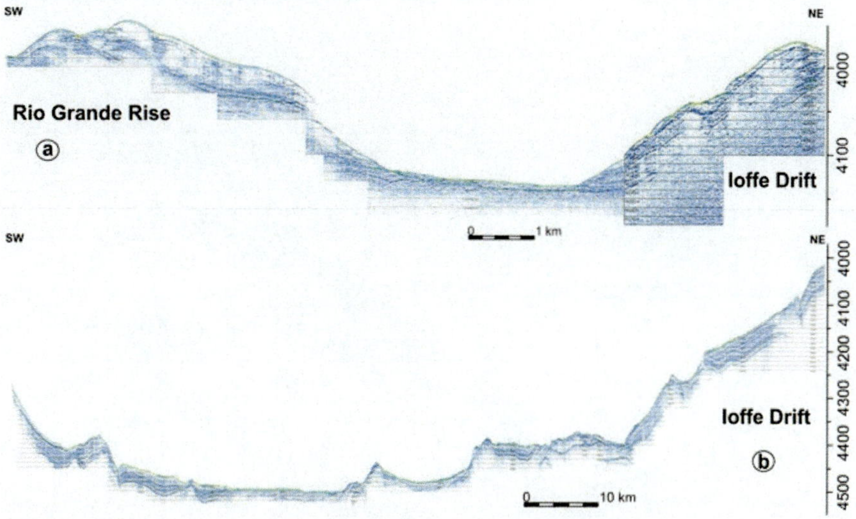

Fig. 9 High-resolution sub-bottom profiles (SES 2000 deep) running from the fault-controlled valley of the Florianopolis Fracture Zone to the Ioffe Drift to the south

slumps (Barker et al. 1983). This explains the minimal sub-bottom penetration in the *SES-2000 deep* profiles. The apparent thickness of the acoustically transparent and sometimes layered deposits (probably weakly consolidated gravitites) in the rare small depressions ranges from a few meters to 15 m (Levchenko et al. 2020). Mass-movement deposits also occur in some places on the slopes.

Possible contourite sediment waves were traced in the middle part of the slope on the southern rise, with a proposed contourite terrace in its uppermost part; however, the interpretation of these features is ambiguous. There appear to be mixed turbidite-contourite deposits at the base of the rise, while clear, distinct contourite features (sediment waves and probably small, sheeted drift) are apparent in the Cruzeiro do Sul Trough (Fig. 10).

4 Conclusions

The Ioffe Drift is an asymmetric lens-like mounded depositional body overlying the central part of a large tectonic ridge extending WSW–ENE within the Florianopolis Fracture Zone. The overall geometry, orientation, erosional and depositional features expressed in the body's morphology, the shape of the seismic units and the internal unconformities indicate that it can be classified as a mounded contourite drift. Its overall geometry is controled by the basement relief and influence of the bottom current (mainly LCDW). Four seismic units, reflecting varying bottom conditions and regimes of sedimentation, were identified in the drift structure. The acoustically

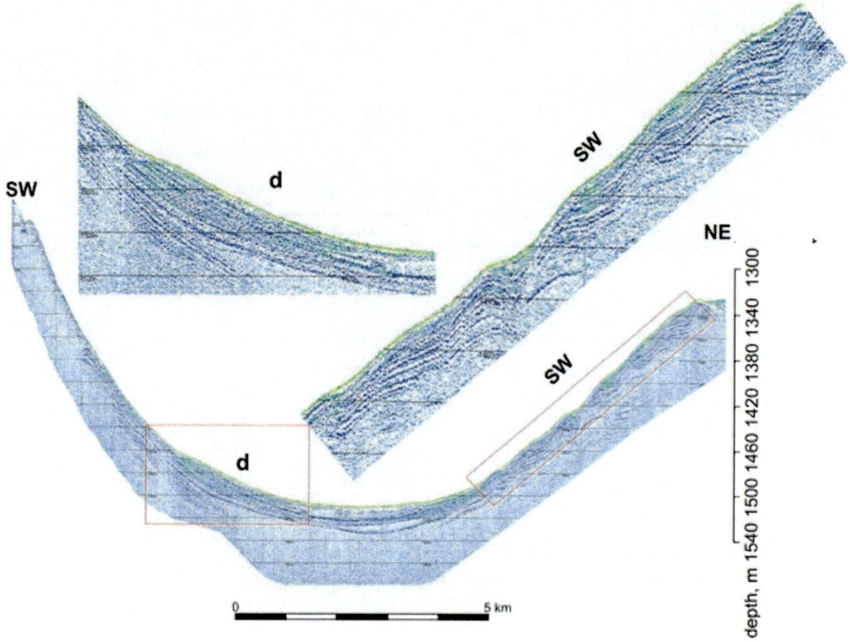

Fig. 10 Contourite features (sediment waves and probably some sheeted drift) in the Cruzeiro do Sul Trough on the top of the Rio Grande Rise; d–drift

transparent unit probably corresponds to tranquil bottom conditions and predominant hemipelagic settling, while the stratified units reflect the strong impact of oceanographic processes on sedimentation. Lateral variations in unit thicknesses demonstrate changes in the location and shape of the main depocenters of the drift, probably corresponding to changes to local bottom-circulation patterns. The seismoacoustic structure of the upper part of the drift is characterized by clear evidence of drastic erosion caused by bottom-current activity.

The correlation sub-bottom profiling data collected on the drift and on the Rio Grande Rise demonstrates the lower impact of bottom currents on sedimentation across the rise.

References

Barker PF, Buffler RT, Gamboa LA (1983) A seismic reflection study of the Rio Grande Rise. In: Barker PF, Johnson DA, Carlson RL (eds) Initial Reports of the Deep Sea Drilling Project, 72. U.S. Government Printing Office, Washington

Dianskii NA, Bagno AV, Zalesny VB (2002) Sigma-model for global ocean circulation and its sensitivity to variations in wind friction stresses. Izv Akad Nauk Fiz Atmos I OKEANA 38:537–556

Faugères J-C, Stow DAV, Imbert P, Viana A (1999) Seismic features diagnostic of contourite drifts. Mar Geol 162:1–38. https://doi.org/10.1016/S0025-3227(99)00068-7

Gamboa LA, Buffler RT, Barker PF (1983) Seismic stratigraphy and geologic history of the Rio Grande Gap and Southern Brazil Basin. In: Barker PF, Johnson DA, Carlson RL (eds) Initial reports of the deep-sea drilling project, 72. Government Printing Office, Washington, U.S, pp 481–496

GEBCO Compilation Group (2020) GEBCO 2020 Grid

Hernández-Molina FJ, Maldonado A, Stow DAV (2008) Abyssal plain contourites. In: Rebesco M, Camerlenghi A (eds) Contourites: developments in sedimentology, 60. Elsevier, Amsterdam, pp 345–378

Ivanova E, Borisov D, Dmitrenko O, Murdmaa I (2020) Hiatuses in the late Pliocene-Pleistocene stratigraphy of the Ioffe calcareous contourite drift, western South Atlantic. Mar Pet Geol 111:624–637. https://doi.org/10.1016/j.marpetgeo.2019.08.031

Ivanova E, Murdmaa I, Borisov D et al (2016) Late Pliocene-Pleistocene stratigraphy and history of formation of the Ioffe calcareous contourite drift, Western South Atlantic. Mar Geol 372:17–30. https://doi.org/10.1016/j.margeo.2015.12.002

Johnson DA (1984) The Vema channel: physiography, structure, and sediment—current interactions. Mar Geol 58:1–34. https://doi.org/10.1016/0025-3227(84)90114-2

Levchenko OV, Borisov DG, Libina NV (2020) Contourites and gravitites on the Rio Grande Rise, southwest Atlantic Ocean (Seismoacoustic Data). Lithol Miner Resour 55:165–176. https://doi.org/10.1134/S0024490220030037

Nielsen T, Knutz PC, Kuijpers A (2008) Seismic expression of contourite depositional systems. In: Rebesco M, Camerlenghi ABT-D in S (eds) Contourites, chapter 16. Elsevier, pp 301–321

Rebesco M, Hernández-Molina FJ, Van Rooij D, Wåhlin A (2014) Contourites and associated sediments controlled by deep-water circulation processes: state-of-the-art and future considerations. Mar Geol 352:111–154. https://doi.org/10.1016/j.margeo.2014.03.011

Rebesco M, Stow D (2001) Seismic expression of contourites and related deposits: a preface. Mar Geophys Res 22:303–308. https://doi.org/10.1023/A:1016316913639

Lithology

Ivar Murdmaa, Dmitrii Borisov, Evgenia Dorokhova, and Olga Dara

Abstract The uppermost seismic unit of the Ioffe Drift partially recovered by five sediment cores is mainly composed of alternating foraminiferal and nanno-foraminiferal high-calcareous oozes with a minor contribution of foram-nannofossil and nannofossil oozes. The major characteristics and possible diagnostic criteria of the Ioffe Drift calcareous contourites were studied on the basis of a detailed investigation of grain-size distribution, color reflectance, magnetic susceptibility and the chemical and mineral composition of the drift sediments. The analysis of this large dataset revealed the main sources of the biogenic sedimentary material and the mechanisms of the material's transport and deposition. The variability patterns of the studied parameters suggest erosion-depositional changes in bottom-current activity. The end-member modeling revealed unusually high contribution of EM2 mainly controlled by the sortable silt (*SS*) content. Most cores show significant correlation between *SS* content in fractions <63 μm and mean *SS* particle size thus confirming the influence of bottom-currents on the sedimentation. Finding of authigenic Mn-oxide minerals in sediments supports the oxic environment of sedimentation and early diagenesis.

I. Murdmaa · D. Borisov (✉) · E. Dorokhova · O. Dara
Shirshov Institute of Oceanology, Russian Academy of Sciences, Moscow, Russia
e-mail: dborisov@ocean.ru

I. Murdmaa
e-mail: murdmaa@mail.ru

O. Dara
e-mail: olgadara@mail.ru

I. Murdmaa and E. Ivanova (eds.), *The Ioffe Drift*, Springer Geology,
https://doi.org/10.1007/978-3-030-82871-4_6

1 Methods

1.1 Visual Description and Smear-Slide Examination

Each core was opened, described and continuously sampled on board. The relative abundance of major sediment-forming components (foraminiferal tests, nannofossils and their fragments) was examined in smear-slides under the polarizing microscope. The results of microscopic study were used for a preliminary identification of the type of sediments throughout the cores.

1.2 Core Logging, $CaCO_3$ and TOC Content

Cores AI-3316, AI-3317, AI-3318, AI-3319 and AI-3655 were subjected to a multi-proxy study using the *Geotek Core Workstation (MSCL-XYZ)*. The study was performed on split cores in automatic mode, and included photo scanning, magnetic susceptibility measurements, spectrophotometry and X-ray fluorescence (XRF) analyses.

The core photos were acquired using a *Geotek Geoscan V* color linescan camera (1 × 5 K pixels CCD matrix) in visible light with polarizing filters. The core imaging was carried out at the highest possible resolution (400 lines/cm, pixel size of 25 μm).

Volume magnetic susceptibility (κ) measurements were performed (AI-3316, AI-3317, AI-3318, AI-3319 and AI-3655) using the *Bartington MS3* system and the *Bartington MS2E* sensor. The down-core measurement resolution was 0.5 cm (1 s measurement time with zeroing after each measurement). The magnetic susceptibility (MS) in core AI-2436 was measured in individual samples in manual mode using the same MS measurement system (2 cm down-core resolution, 5 s measurement time with zeroing after each measurement). The results were reported in SI 10^{-5} units.

Sediment reflectance spectra over the visible domain (400–700 nm, wavelength pitch of 10 nm) were measured on split cores (AI-3316, AI-3317, AI-3318, AI-3319 and AI-3655) using the *Konica Minolta CM-700d* spectrophotometer. The measurements were carried out in SCE mode (specular component excluded), with an 8 mm diameter of measurement area, observer conditions of 2° and D65 illuminant condition, corresponding to the correlated color temperature of 6,504 K. The color of sediments was estimated using the CIE L*a*b* color space. This three-dimensional space approximates a cylinder with a vertical axis corresponding to L (lightness, similar to the Munsell value). Parameters a* and b* represent Cartesian coordinates defining a quasi-circular hue-chroma space at any given L* value. This color space is widely used for color estimation in ODP/IODP (Saito 1996; Balsam et al. 1997; Giosan et al. 2002).

The elemental composition of sediments was studied using the ultra-sensitive *Geotek XRF spectrometer* (15 W/50 kV, Rh anode) with a helium-flushed measurement cell, with a measurement area of 15 × 10 mm, 1 cm down-core resolution and 1 s live time for each beam. The analysis was carried out using 10 kV (70μA) and 40 kV (95μA, with 125 μm Ag filter) tube settings. The XRF spectral data were processed using *bAxil* software and the results presented as ratios of element concentrations (counts per second) or as normalized values. Peak-area element profiles were normalized by the total scatter (incoherent + coherent scatter) to minimize the effects of the water content (Guyard et al. 2007; Kylander et al. 2011; Berntsson et al. 2014; Shala et al. 2014). The split cores for all types of measurements were covered with 416 Prolene 4 μm thin film.

Total organic carbon (TOC) and calcium carbonate content ($CaCO_3$) were measured using an *AN-7529* express coulometric carbon analyzer (Gomel Plant of Measuring Equipment, Republic of Belarus) on 195 dried sediment samples (about 0.25 g each) taken at 2–10 cm intervals from core AI-2436 and at 10–20 cm intervals from the others. The samples were dried at a temperature of 100 °C. The carbonate content was calculated from the CO_2 content.

1.3 Grain-Size Analysis

Grain-size analyses were carried out using a *SALD-2300* laser diffraction particle-size analyzer (Shimadzu, Japan) at the Laboratory of the Atlantic Geology, Atlantic Branch of the Shirshov Institute of Oceanology, Russian Academy of Sciences (Kaliningrad, Russia). *SALD-2300* covers a size spectrum from 17 to 2.5 mm. Before the analysis, particle disaggregation was carried out using sodium tripolyphosphate and ultrasonic bath on all samples. The grain-size distribution was studied in 48 samples from core AI-3316, 46 samples from AI-3317, 43 samples from AI-3318, 47 samples from AI-3319, 61 samples from AI-3655 and 76 samples from AI-2436.

Grain-size fractions were determined following the modified scale by Wentworth (1922), where the boundaries between sand and silt and between silt and clay correspond to 63 μm and 2 μm, respectively (Friedman and Sanders 1978). The textural name for the sediment was allocated according to the classification by Shepard (1954).

The end-member modeling approach was applied to unmix the original grain-size distribution (GSD) on grain-size subpopulations, which reflects sediment provenance and/or transport mechanism (Weltje 1997; Weltje and Prins 2007). The mathematically estimated end-members allow an evaluation of the relative contribution of the main sedimentary material sources or processes of sediment transport into the initial GSD and to reconstruct sedimentation in various environments (Prins et al. 2002; Weltje and Prins 2003; Hamann et al. 2008; IJmker et al. 2012; Just et al. 2012). The non-calcareous (terrigenous) fraction of the sediment is traditionally used in end-member modeling. In our case, to study the sedimentation environment of the calcareous contourite drift we used bulk samples consisting almost totally of $CaCO_3$ particles except for clay-size fraction (<2 μm) that contains terrigenous clay.

The end-member modeling algorithm was developed by Paterson and Heslop (2015). The data were processed using the Matlab-based software, AnalySize

1.1.2 (Martín-Chivelet et al. 2008). Non-parametric unmixing was carried out. The minimum number of end-members estimated by goodness-of-fit statistics appears to be three for the entire set of analyzed samples. This number adequately describes the initial GSD at each site. Correlation of the sortable silt (*SS*) content in the sum of size fractions <63 μm and the mean size of *S* particles is calculated for all cores to judge about the bottom-currents influence on sedimentation (according to McCave et al. 2017).

1.4 Mineralogical Analysis and Scanning Electron Microscopy

The mineral composition was analyzed using the X-ray diffractometer *D8 ADVANCE* (Bruker AXC), Cu-Kα, with a Ni 0.02 filter, 40 kV, 40 mA and the linear detector *LYNXEYE*. Scanning was performed in a discrete mode with a step of 0.02° and an exposure of 4 s/step in the range from 2.5° to 70° 2ϴ, with rotation. DIFFRAC.EVA software was applied for primary processing, interpretation of spectra and calculation. Mass quantitative analysis was performed using corundum numbers from the ICDD database. Special cuvettes, a variety of optical shutters for background scattering made from a single silicon crystal, were used for this precision work. Before the XRD (X-ray Diffraction) analysis, each sample was studied under a *Stemi 508* binocular microscope (Carl Zeiss Microscopy GmbH, Germany).

The identification of clay minerals was carried out on oriented air-dried preparations from a suspension of the clay fraction separated from the sample in distilled water. Then the preparations were saturated with ethylene glycol (to detect the minerals of the smectite group). If needed, the samples were heated at 550 °C to test for kaolinite and chlorite (Brown 1961; Moore and Reynolds 1997).

The sediment samples from the core intervals with dark-colored grains were washed in distilled water to remove clay minerals and then treated with hydrochloric acid to dissolve $CaCO_3$ and to extract other minerals. The resulting material was photographed and studied by powder X-ray diffraction. The diagnostics of manganese minerals were carried out in two stages using the method suggested by O.G. Smetannikova (Smetannikova 1988). Analysis of the sediment composition in the fractions 63–100 and >100 μm was performed on sediment samples from three intervals of core AI-2436, using a *LEO Supra 50 VP* field emission scanning electron microscope (Carl Zeiss, Germany) with an accelerating voltage of 25 kV in the low-vacuum nitrogen atmosphere mode (40 Pa).

2 Lithology of Core Sections

The cores lithology is mainly based on variations in grain-size distribution, while the chemical (mainly $CaCO_3$ content) and mineral composition (apart from dominating calcite) provides further information about the sediments recovered in cores from the uppermost parallel-stratified seismic facies. Core logging, color reflectance and MS measurements have been made to distinguish sedimentary structures and hiatuses. The cores description is based on our classification that allows to distinguish the biogenic genetic types of sediments using grain-size distribution data (proportion of two major biogenic components: foraminiferal tests and calcareous nannofossils).

2.1 *Classification of Recovered Sediments*

Calcareous oozes and rocks are commonly classified by their calcium carbonate content. A modified quartile scale of $CaCO_3$ was used to distinguish the sediment types in the Ioffe Drift. All sediments recovered from the Ioffe Drift cores (see Fig. 1b and Table 1 in Chap. 4) are of high-calcareous ooze (>75% $CaCO_3$), apart from the rare samples with $CaCO_3$ ranging between 70 and 75% thus formally falling into the clayey calcareous ooze type. Two subtypes of high-calcareous ooze were distinguished: very high-calcareous ooze (>85% $CaCO_3$) and slightly clayey high-calcareous ooze (75–85%). Both form layers within the upper Ioffe Drift.

Sediments with a lower $CaCO_3$ content occur only in core AI-3319, retrieved to the WSW of the drift at a water depth of 4,066 m, well below the present lysocline. They include the following sediment types, distinguished by quartile $CaCO_3$ content: clayey calcareous ooze (50–75% $CaCO_3$); and calcareous clayey mud (25–50% $CaCO_3$).

Planktic foraminiferal tests and nannofossils strongly dominate among the biogenic source material in the sediments of the drift. The $CaCO_3$ content is represented both by intact initial biomorphic particles and their fragments of various sizes. We attempted to classify the calcareous oozes according to our estimation of the approximate relationship between the foraminiferal and calcareous nannofossil materials, using the measured GSD of the high-calcareous oozes. The sand fraction (>63 μm) consists exclusively of intact and fragmented foraminiferal tests. The size of most nannofossils does not exceed 10 μm, the smallest sortable silt fraction size limit. Some rather rare large nannofossils (with a diameter up to 30 μm) may occur in the *SS* fraction (10–63 μm), yet all abundant sediment-forming Quaternary and Pliocene nannofossil species are less than 10 μm in diameter. So, the sum of sand and *SS* fraction—that is, the biogenic calcareous particles coarser than 10 μm—can be taken as a minimum measure of foraminiferal component in the ooze. Note that terrigenous particles of this size, including *SS*, are not found in the sediments from the Ioffe Drift.

Intact calcareous nannofossils (coccoliths and other groups of very thin calcareous plates covering coccolithophore cells) and their determinable fragments are concentrated in the fine silt fraction (2–10 μm), together with an uncertain amount of fine foraminiferal fragments. Clay-size crystallites (micrite) of both foraminiferal and nannofossil origin probably occur in the finest fraction <2 μm, but these cannot be identified directly. To eliminate the influence of this uncertainty, the <2 μm fraction was excluded from the total GSD used in the genetic classification (Table 1).

Table 1 Minimum, maximum and mean values of parameters characterizing the grain-size of the Ioffe Drift sediments

	AI-2436	AI-3316	AI-3317	AI-3318	AI-3319	AI-3655
No. of samples	76	48	46	43	49	61
Core length, cm	714	489	515	346	287	515
Core length/samples	9.4	10.2	11.2	8.0	5.9	8.4
Sand, %						
Min	1.78	0.15	0.00	1.94	0.00	10.30
Max	69.18	77.40	94.88	88.79	61.79	57.25
Mean	36.18	24.00	28.43	38.54	13.31	35.50
Sortable silt (SS), %						
Min	21.31	15.26	1.49	8.30	12.29	18.70
Max	54.49	52.89	45.88	44.77	46.26	36.00
Mean	39.52	36.56	26.97	32.20	32.35	25.04
Fine silt, %						
Min	3.07	0.04	3.63	1.86	12.98	9.87
Max	31.64	61.60	64.26	39.21	58.70	42.06
Mean	13.45	26.10	29.59	18.75	35.69	26.73
Clay, %						
Min	0.00	0.00	0.00	0.00	0.00	5.73
Max	34.93	23.45	32.71	24.59	35.00	22.05
Mean	10.85	13.34	15.02	10.50	18.65	12.74
Sorting						
Min	2.60	2.04	1.92	1.63	2.62	4.26
Max	7.20	7.09	6.81	6.65	6.30	7.19
Mean	4.56	4.83	5.07	4.36	4.45	5.89
Mean size, mkm						
Min	5.45	3.76	3.00	3.88	2.93	8.37
Max	113.05	100.01	250.48	219.82	83.20	50.59
Mean	36.78	21.27	25.21	42.05	12.37	21.97

(continued)

Table 1 (continued)

	AI-2436	AI-3316	AI-3317	AI-3318	AI-3319	AI-3655
Forams, %						
Min	52.73	20.01	7.53	51.28	17.61	47.27
Max	96.87	99.96	96.37	98.14	87.02	89.53
Mean	84.19	69.09	63.34	78.23	54.55	69.00
Nanno, %						
Min	3.13	0.04	3.63	1.86	12.98	10.47
Max	47.27	79.99	92.47	48.72	82.39	52.73
Mean	15.81	30.91	36.66	21.77	45.45	31.00
Foram/nnanno ratio						
Min	1.12	0.25	0.08	1.05	0.21	0.90
Max	30.98	17.21	26.57	52.77	6.71	8.55
Mean	8.01	3.15	3.19	7.67	1.67	2.67

Thus, relationship between two grain-size components, termed the foraminiferal (>10 μm) and the nannofossil (2–10 μm), were used to classify the biogenic calcareous material. Four quartile genetic types may be distinguished, with boundaries at 25, 50 and 75% of each component: foraminiferal; nanno-foraminiferal; foram-nannofossil; and nannofossil oozes. Additional subtypes, such as pure foraminiferal sand (more than 90% calcareous sand) and pure nannofossil ooze (more than 90% calcareous fine silt) were also distinguished to characterize the rather rare yet genetically important thin interlayers. Note that this preliminary classification is valid only for the high-calcareous oozes that dominate strongly in cores from the Ioffe Drift itself, not for its deep margins below the foraminiferal lysocline, which are affected by strong calcium carbonate dissolution. The grain-size distribution of the calcareous biogenic material in sediments of a lower $CaCO_3$ content from the base of the ridge covered by the Ioffe Drift (e.g., core AI-3319) should follow the same order, as it also consists of foraminiferal and nannofossil calcareous remains, which mainly have the same grain-size limits. The non-calcareous terrigenous material in these sediments is very fine-grained and may affect only the finest (<2 μm) fraction, which is ignored in our classification.

2.2 *Lithostratigraphy of Individual Cores*

Core AI-2436 (Fig. 1) was the first collected from the contourite Ioffe Drift in 2010. The core was retrieved at a water depth of 3,800 m on the drift summit (see Figs. 1b and 5 and Table 1 in Chaps. 4 and 5, an unknown portion was lost from the core top, probably comprising fluidized foraminiferal sand. We arbitrarily assumed a core

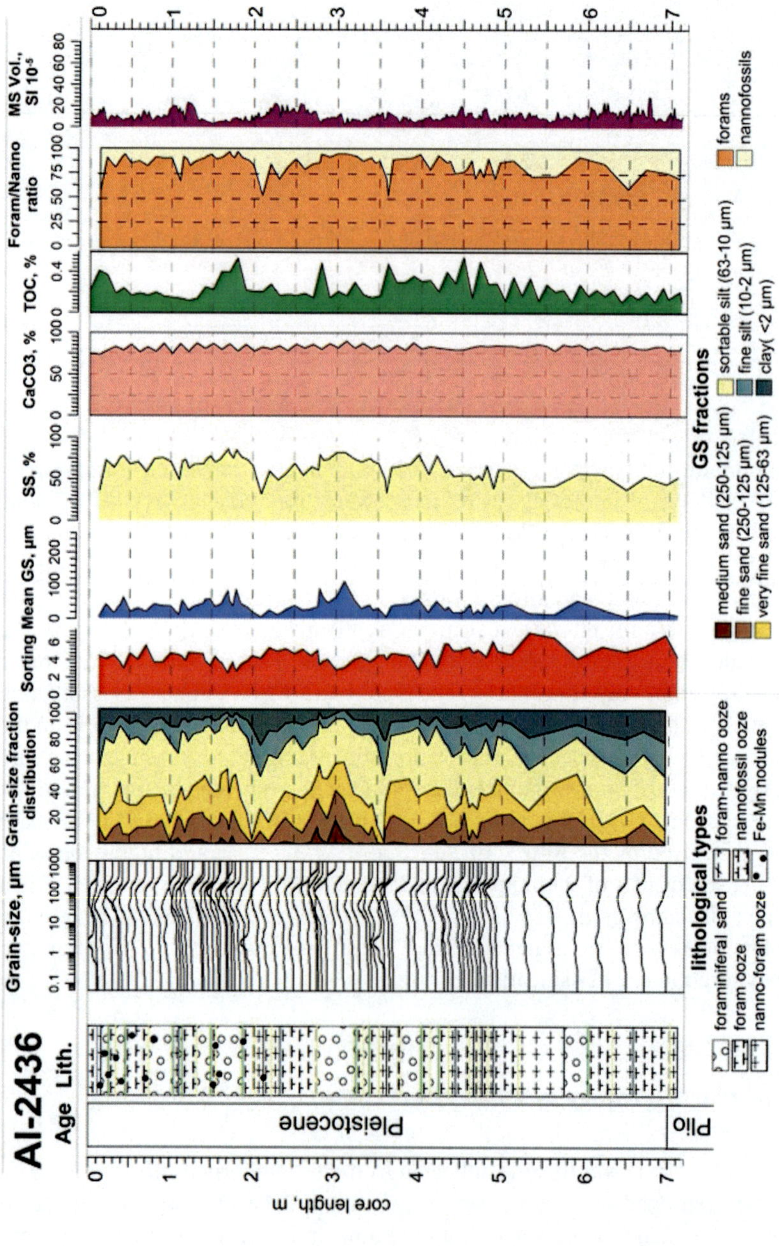

Fig. 1 Generalized age, lithostratigraphy of sediment types, grain-size distribution parameters (GS), $CaCO_3$, TOC content and volume magnetic susceptibility (MS Vol.) of core AI-2436

length of 7.14 m (Ivanova et al. 2016), and all intervals in the following figures and description follow this assumption.

Core AI-2436 recovered the entire Quaternary section and likely entered into the Upper Pliocene deposits below (Ivanova et al. 2016, 2020; see Chap. 7). The stratigraphic sequence is interrupted by numerous long- and short-term hiatuses that span a significant part of the geological time scale, especially before the last 0.27 Ma when sedimentation became likely more continuous (see Chaps. 7 and 9).

The Upper Pliocene and Lower Pleistocene deposits are represented by layers of stiff calcareous ooze separated by thin interlayers of foraminiferal sand, possibly artificially fluidized during the core extraction from the tube. They likely mark hiatus surfaces.

The core consists entirely of light-colored (10YR 8/2, 10YR 7/4) high-calcareous ooze with a stepwise increase in density down-core, from a soupy consistence in the uppermost layer of unknown thickness to a stiff, slightly lithified sediment at the bottom. The physical properties were not measured and are evaluated only by shipboard visual core description.

Sedimentary structures include bioturbation and indistinct layering, expressed mainly in alternating decimeter-scale layers of varying mean diameters, content of sand-size foraminiferal tests, *SS*, sand fractions, and thus, foraminiferal component (Fig. 1). Bioturbation, common in contourites (Stow and Faugères 2008; Wetzel et al. 2008; Rebesco et al. 2014), is best expressed at the layer boundaries. Slightly clayey (73–80% $CaCO_3$), darker-colored layers occur in the Upper Pleistocene deposits.

Authigenic (likely diagenetic) sand-size calcite microdruses occur in the stiff ooze from the lower part of the core. The Upper-Middle Pleistocene sediments contain ferromanganese nodules, described in Chap. 8 and shown in Figs. 1, (Figs. 1 and 2 in Chap. 8).

Variations in the main GSD parameters are shown in Fig. 1. Diagrams of continuous GSDs for each analyzed sample demonstrate common bimodality or polymodality among the fine fractions, along with distinct high modes within the sand fractions. This probably reflects contrasting transport and settling mechanisms affected by variations in bottom-currents velocities, as described in Sect. 3.1.

The lower part of the core AI-2436, 714–515 cm below sea floor (bsf) is represented by the fine-grained nanno-foraminiferal calcareous ooze with a sharply coarser foraminiferal ooze interlayer at 570–540 cm (Fig. 1). The relatively fine-grained two-meter-long layer corresponds approximately to the Upper Pliocene–Middle Pleistocene (see Chap. 7). It is characterized by rather low sand, relatively high silt and somewhat increased clay content in this core. Together with *SS* prevailing over fine silt, this means the domination of the foraminiferal component over the nannofossil one, although the latter is somewhat higher than in the coarser interlayer. The sediment is very poorly sorted. The coarse interlayer identified by two grain-size analyses at 570 cm and 540 cm possibly represents a coarse-grained deposit fragment enriched in foraminiferal sand between hiatuses.

The next interval (about 515–370 cm) was identified on the basis of its low-amplitude saw-shaped oscillations at a somewhat higher mean diameter and sand content level. It likely corresponds to the Middle Pleistocene (see Chap. 7), with its long-term hiatuses (see Chap. 9). The *SS* content and foraminiferal component oscillate at a slightly higher level than in the interval below, probably reflecting a more intense and selective reworking of the fine-silt and clay-size fractions. Sorting is somewhat better, but still poor. The high-calcareous sediment is classified as foraminiferal ooze with a minor admixture of the nannofossil component.

The unusually thick Upper Pleistocene deposit, relative to the other cores, differs from the sediments below in its high-amplitude changes of GSD parameters involving maximum and minimum values. Its lower boundary is marked by a distinct thin, fine-grained interlayer dominated by fine silt and thus by nannofossils. Further upward, two bi-gradational contourite cyclites are distinguished with coarse fraction peaks in their centers (Fig. 1) indicating gradational increase and decrease of bottom-current velocities. They are separated by fine-grained ooze suggesting slowing down the bottom-current at this site. A foraminiferal sand layer occurs at the core top deformed during the core extraction.

Similar thin and sharp interlayers also occur in the section above (Fig. 1). They are followed by thicker layers of much coarser mean diameter and high-foraminiferal sand content with sharp peaks in the middle parts resembling the bi-gradation typical of contourite sedimentary structures (Stow and Faugères 2008; Stow et al. 2019). The high *SS* content (up to 54%) supports the suggestion of a strong bottom-current influence on the coarsest sediment deposition. If so, the fine-grained interlayers mark still-water boundaries between the contourite cyclites. This suggests rather continuous contourite deposition without considerable hiatuses. The Upper Pleistocene section contains ferromanganese nodules especially abundant at 16–82 cm, which are described in Chap. 8.

The comparison to core AI-3655 (see below), which was retrieved very close to the AI-2436 site, demonstrates quite different facies. First of all, there is a much shorter simultaneous section, thus expressing lower average sedimentation rates and an absence of distinct contourite cycles. This possibly reflects the occurrence of hiatuses at AI-3655 rather than any gravity-flow activity at AI-2436, as we thought earlier.

The Upper Pleistocene deposits show mainly poor sorting, but the sorting coefficient decreases somewhat in the coarse mean diameter peaks (Fig. 1) and, together with the high *SS* content, this supports the suggestion of better sorting due to stronger bottom-current activity.

Core AI-3316 (Fig. 2 and Plate 1) was retrieved from the gentle upper northeastern slope of the Ioffe Drift at a water depth of 3,900 m (see Table 1in Chap. 4). It went through the parallel-stratified uppermost seismic unit (see Figs. 4a and 5 in Chap. 5) to penetrate the entire relatively thin (250–300 cm long, depending on the rather uncertain depth of the Pliocene–Pleistocene boundary) Quaternary section and recover about 240 cm of the Upper Pliocene deposits below with an age of about 3 Ma (see Chap. 7).

Both the Pliocene and the Quaternary sediments in the core are represented by high-calcareous ooze (Fig. 2). However, the background $CaCO_3$ content is slightly lower than that in cores taken at shallower water depths possibly owing to a partial dissolution of foraminifers close to the lysocline (see Chap. 7).

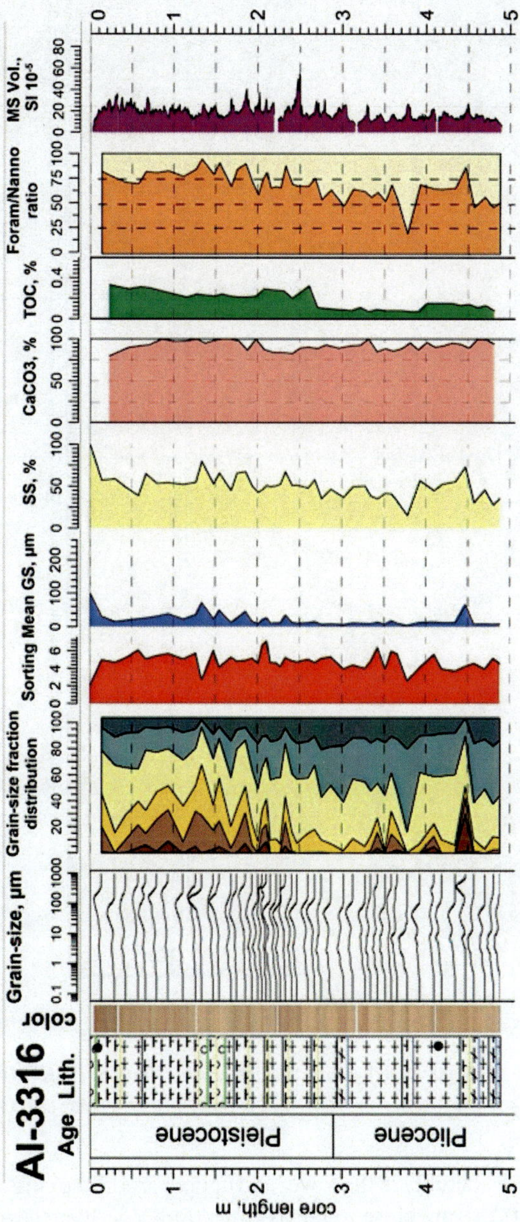

Fig. 2 Generalized age, lithostratigraphy of sediment types, color, grain-size distribution parameters (GS), $CaCO_3$, TOC content and volume magnetic susceptibility (MS Vol.) of core. For symbols, see Fig. 1

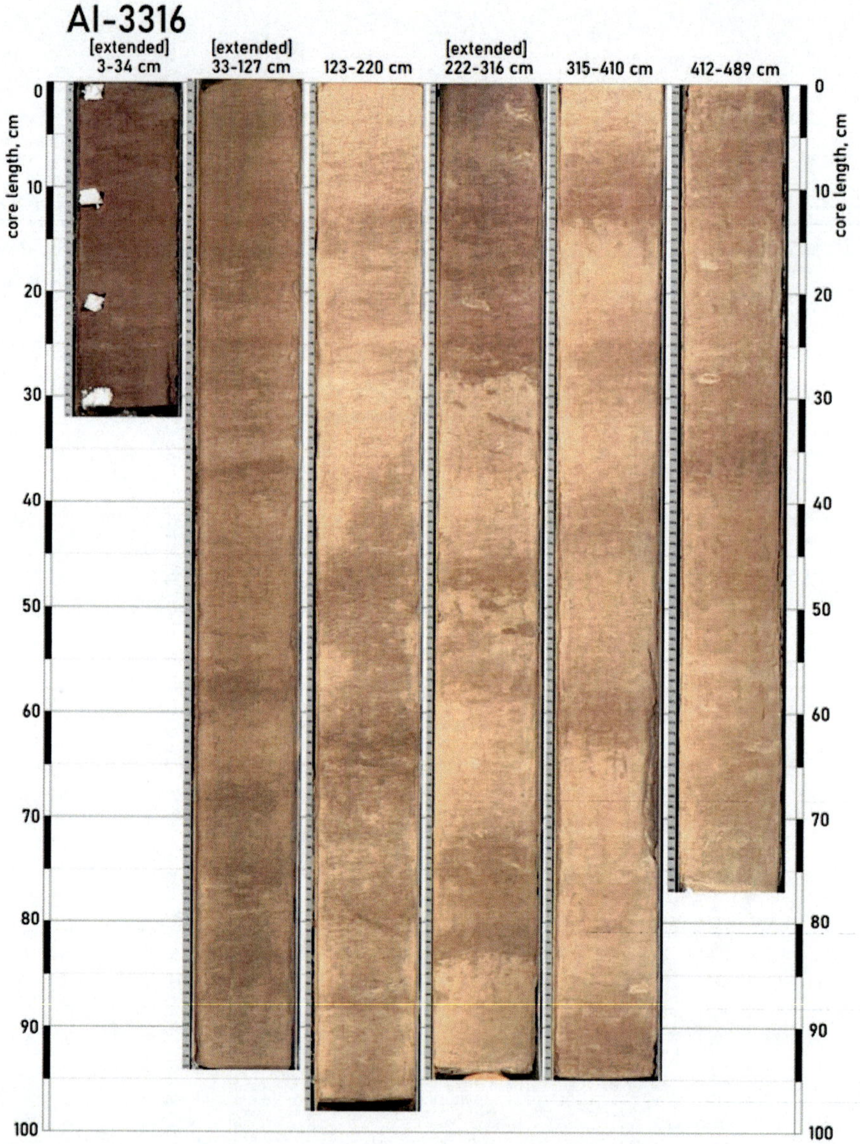

Plate 1 Photo of core AI-3316. [extended] means that the core section was slightly extended during storage

The GSD differs considerably in the lower and upper parts of the core section, with the transition at about 250 cm, close to the assumed Pliocene/Pleistocene boundary (see Chap. 7). The Upper Pliocene deposits are finer grained and are notably rather well sorted, whereas the sand interlayer demonstrates poor sorting. The *SS* content of the background sediment is moderately low and decreases further in the fine-grained interlayer. According to our classification the sediments below about 250 cm bsf are represented by nanno-foraminiferal ooze with minor interlayers of foram, foram-nannofossil and nannofossil ooze.

The transition to coarser Pleistocene deposits with a mean diameter ranging from 5.4 to 15 μm takes place through moderate-amplitude oscillations to a peak at 233 cm. Further up-core, mean particle size decreases to its minimum value, corresponding to low sand content and unusually high fine silt content, indicating an increased nannofossil proportion. The *SS* content shows two peaks with a maximum value of 53%. A sharp increase was documented in the sand content, corresponding to the maximum value of mean particle size up to 100 μm at the core top, with a well-sorted foraminiferal sand (mean grain size 80 μm) (Fig. 2). It contains an admixture of *SS* and no finer fractions, which were apparently removed by strong bottom-currents. Sorting is poor, with the bimodal or polymodal GSD in turn suggesting a mixture of particles from contrasting transport mechanisms. The sediment is classified as foraminiferal ooze.

Core AI-3317 (Fig. 3 and Plate 2) is from a small terrace on the Ioffe Drift's northern slope at a water depth of 3,832 m (see Fig. 5b in Chap. 5). As shown by its seismic profile, the site comprises a small plateau isolated from the major near-bottom coarse material fluxes by a gentle depression, thus it represents a rather independent sedimentary mini-system.

The core recovered about 300 cm of Upper Pliocene deposits, possibly the oldest in our cores (to 4 Ma, see Chap. 7), overlain by about 200 cm of Pleistocene cover. It consists of the finest-grained sediments and thus has the highest nannofossil constituent. The mean particle size is very low throughout (3.76–100 μm, averaging 21.27 μm) apart from the core-top foraminiferal sand layer and a thin foraminiferal interlayer at 19 cm. Both foraminiferal sands are well sorted.

The lowermost layer, below about 400 cm bsf (420 cm according to the magnetic susceptibility record), has a sharp upper boundary, and differs distinctly from the overlain sediments in terms of its very fine grain size and good sorting (Fig. 3). It is represented by the dominating fine silt with a minor content of sortable silt and abundant clay-size particles. According to our classification, this is a nannofossil ooze with a minor foraminiferal constituent. The deposition of a minimum meter-thick, well-sorted nannofossil ooze in the Upper Pliocene at this specific site, partially isolated from the major near-bottom foraminiferal material fluxes, needs further study, in addition to the causes of its sharp upper boundary, which possibly relates to a hiatus (see Chap. 9).

Above this sharp boundary, the Upper Pliocene and the most Quaternary deposits (except for the abovementioned foraminiferal sands) are represented by alternatng nanno-foraminiferal ooze with minor foraminiferal ooze interlayers.

Core AI-3318 (Fig. 4, and Plate 3) was retreived from the Ioffe Drift summit at a water depth of 3,788 m, and may well serve as an example of typical drift growth by bottom-currents (see Fig. 5a in Chap. 5). The core recovered the entire Pleistocene deposits and entered into the Upper Pliocene with its lower part (likely below 250 cm bsf, see Chap. 7). The lower fine-grained ooze layer (about 350–250 cm) demonstrates

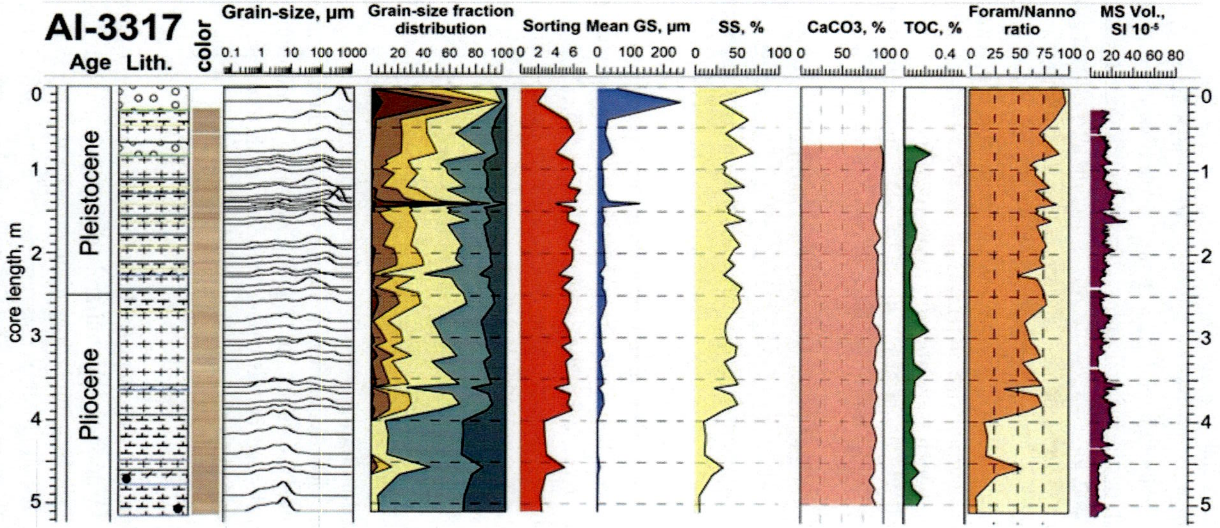

Fig. 3 Generalized age, lithostratigraphy of sediment types, color, grain-size distribution parameters (GS), $CaCO_3$, TOC content and volume magnetic susceptibility (MS Vol.) of core AI-3317. For symbols, see Fig. 1

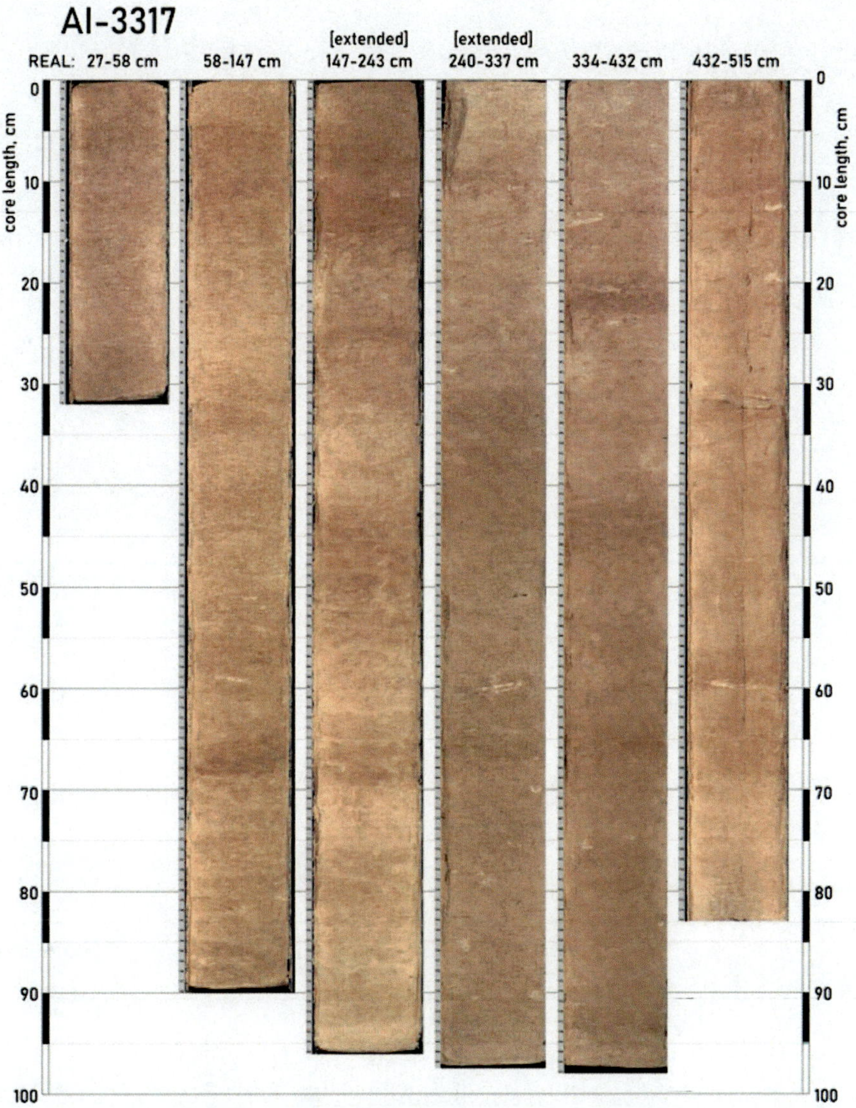

Plate 2 Photo of core AI-3317. [extended] means that the core section was slightly extended during storage

lithological features characteristic of Upper Pliocene deposits in other cores from the drift summit. The mean grain size (42 μm) and *SS* content 32.20% in the core are rather high. The proportion of the nannofossil component relative to the foraminiferal one in the lower part of the core (below 250 cm) is greater than that in the deposits above and the sediment is classified as nanno-foraminiferal ooze.

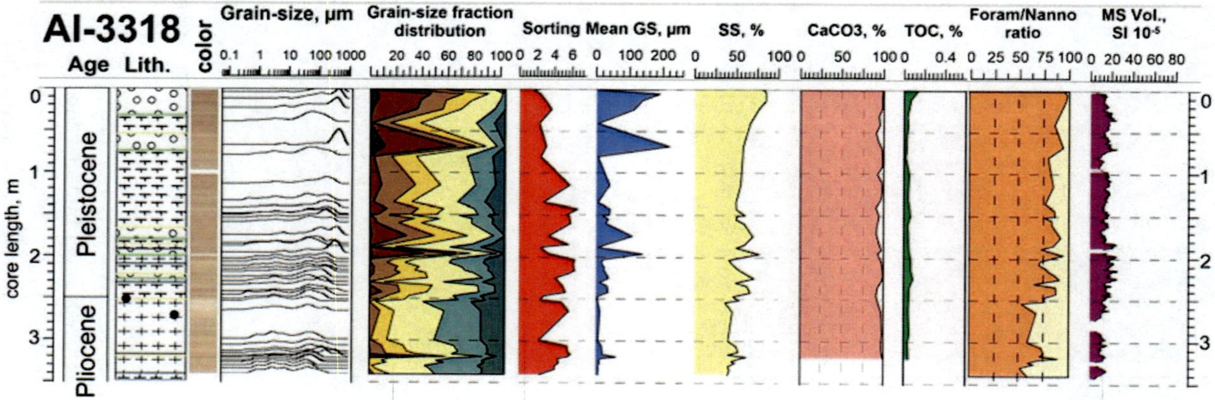

Fig. 4 Generalized age, lithostratigraphy of sediment types, color, grain-size distribution parameters (GS), $CaCO_3$, TOC content and volume magnetic susceptibility (MS Vol.) of core AI-3318. For symbols, see Fig. 1

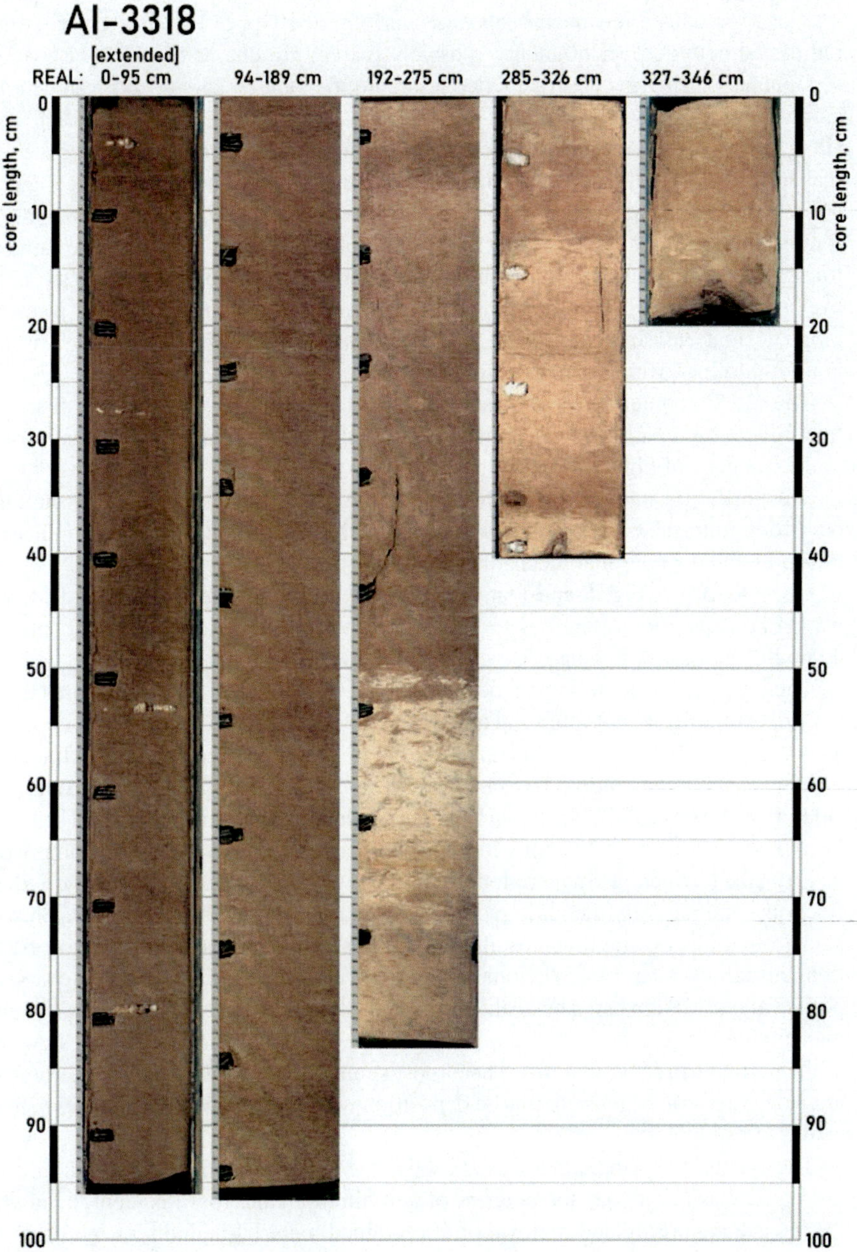

Plate 3 Photo of core AI-3318. [extended] means that the core section was slightly extended during storage

The apparently Pleistocene deposits overlying the Upper Pliocene layer with a rather distinctive upper boundary (possibly hiatus) are characterized by a layered sedimentary structure with a moderate-amplitude (about 250–80 cm) and high-amplitude (above 80 cm to core top) alternation of foraminiferal sand-rich and silt-rich layers. The rare and uneven sampling for grain-size analyses prevents any realistic estimate of the proportion of foraminiferal sand-layer and silt-layer contents. Therefore, the visual image (Fig. 4) may well overestimate the relative amount of foraminiferal sand layers owing to linear interpolation of rare analyses. The same is true of the mean grain-size diagram. The foraminiferal component prevails strongly over the nannofossil, confirming that there was an intense reworking of primary biogenic material and removal of nannofossils and clay-size particles by strong bottom-currents, which increased towards the core top.

The $CaCO_3$ content does not respond markedly to changes in grain-size distribution, reflecting an absence of significant dissolutiom. All analyzed samples belong to the category of high-calcareous oozes.

The sorting is unusually variable. Well and poorly sorted oozes occur among both foraminiferal sand layers and finer-grained intervals, suggesting complicated transport and depositional mechanisms.

Core AI-3319 (Fig. 5 and Plate 4) was collected from the area covered by the wave-like depositional features to WSW from the Ioffe Drift, at a water depth of 4,066 m (see Fig. 5c in Chap. 5).

The core recovered only Pleistocene deposits (Figs. 5 and 12). Unlike cores from the drift summit, it recovered lower calcareous oozes and calcareous clayey mud. This corresponds to the site location well below the present regional lysocline and is mainly related to a partial biogenic $CaCO_3$ dissolution by the agressive Antarctic Bottom Water (AABW). As shown by smear-slide examination and regional settings (see Chap. 2), the non-calcareous component of core sediments is entirely comprised of clay-size terrigenous manerial, clay minerals and finely dispersed quartz. Other fractions, and partially the clay fraction also, consist of biogenic $CaCO_3$. Against a background of low-amplitude oscillations, the mean grain size and grain-size fractions reveal a typical bi-gradational contourite sequence (cyclite) with its coarsest peak at about 120 cm bsf. The cyclite probably reflects an increase in bottom-current intensity on the foraminiferal sand peak (reverse gradation), followed by a slowing of the current (direct gradation). The almost symmetrical cyclite may be interpreted as evidence of continuous contourite deposition without considerable hiatuses (Stow et al. 2019).

The cyclite is also indistinctly expressed in the relationship between foraminiferal and nannofossil components, in terms of a minimum nannofossil content, owing to the intense reworking and removal of fine-grained material including nannofossils. As a whole, the nannofossil component relative to foraminiferal one is rather high, ranging from 12.98 to 82.39% and averaging 45.45% (Table 1).

The sediments are poorly sorted throughout, apart from the fine-grained core-bottom layer and some fine-grained interlayers above. Better sorting of fine-grained, nannofossil-rich layers may be interpreted as the residual nature of both the nannofossil component and the terrigenous clay. Nannofossils resistant to dissolution,

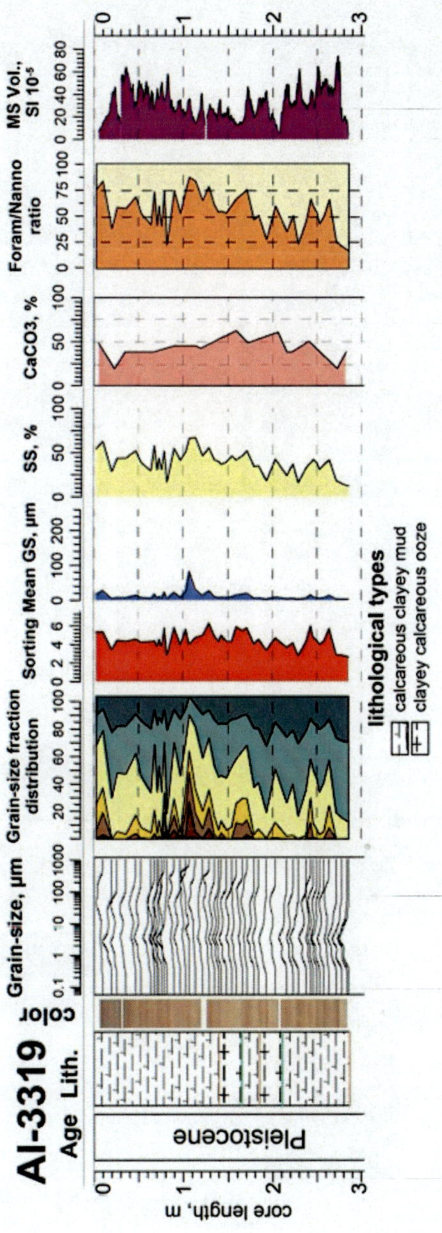

Fig. 5 Generalized age, lithostratigraphy of sediment types, color, grain-size distribution parameters (GS), $CaCO_3$ content and volume magnetic susceptibility (MS Vol.) of core AI-3319

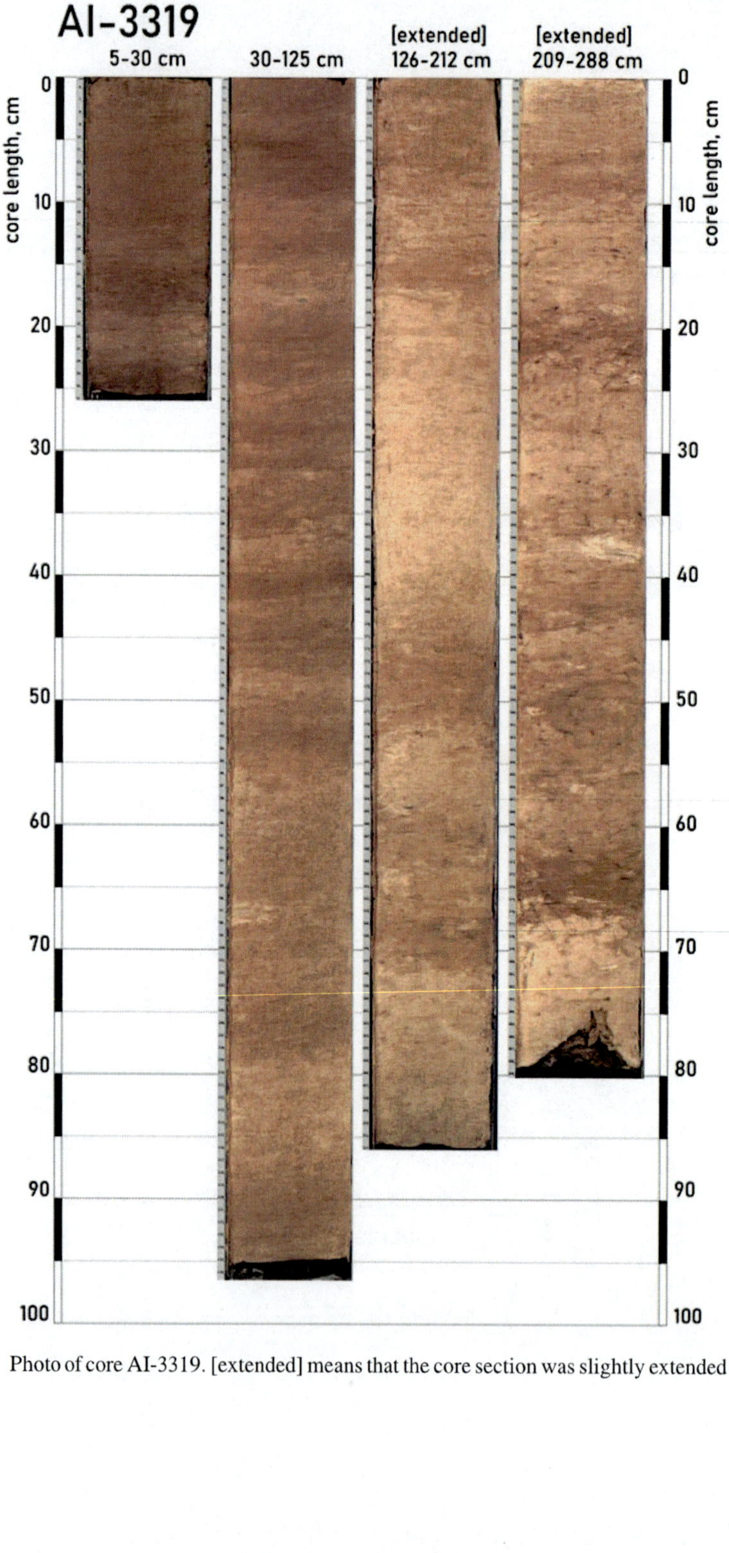

Plate 4 Photo of core AI-3319. [extended] means that the core section was slightly extended during storage

together with the terrigenous clay partially released from coarser foraminiferal component, provide better-sorted material that might improve sediment sorting.

Core AI-3655 was retreived from the Ioffe Drift summit very close to core AI-2436 (see Fig. 5a in Chap. 5) in order to supply the deficit caused by the artificial deformation of a section of this long core; however, it appeared that the new core AI-3655 recovered a principally dissimilar section (Fig. 6 and Plate 5), suggesting an unexpectedly sharp facies change at an indeterminate short distance away and at the same water depth of 3,799 m.

The core penetrated the entire Pleistocene section and entered Upper Pliocene deposits, like AI-2436 discussed above. The Lower and Middle Pleistocene sections are interrupted by long-term hiatuses (see Chap. 9) showing low-amplitude oscillations in sand and silt fractions with no distinct trends. Corresponding irregular variations occur in mean grain size and in the relationship between the foraminiferal and nannofossil components, according to our classification. The sediments are classified as nanno-foraminiferal ooze up to 80 cm bsf and as foraminifreal ooze above.

The main differences between the lithology of cores AI-3655 and AI-2436 appear in the sediments younger than 0.81 Ma. The upper lithologically distinctive layer in AI-3655 (230–0 cm) is represented by a sandy silt with an upward coarsening trend from silt to sandy silt at the core top. In terms of our classification, this means a transition from nanno-foraminiferal ooze to foraminiferal ooze, with a minor admixture of the nannofossil component that suggests a gradual increase in bottom-current intensity.

As shown above, the synchronous interval in core AI-2436 is much thicker and possibly contains several bi-gradational contourite cyclites which are absent in core AI-3655. We hypothesize that the observed sharp facies change might result from the location of the AI-2436 site close to a rather stable suspension-containing bottom-current stream that maintained contiguous contourite deposition. By contrast, the AI-3655 site was probably aside of this active depositional stream. It is also possible that a higher velocity erosional stream occurred over the AI-3655 site resulting in hiatuses which restricted the average sedimentation rates of the Upper Quaternary deposits compared to rapid accumulation at AI-2436 site.

3 Grain-Size Distribution

3.1 Common Features

High-calcareous contourites of the Ioffe Drift recovered in five sediment cores from its upper part are dominated by a relatively coarse grain-size distribution, as compared to common pelagic calcareous oozes of vertical sedimentation. According to the above classification (see Sect. 2.1), this means that foraminiferal source material commonly prevails over fine-grained nannofossil material, although both are initially derived from plankton production in surface waters (zooplankton and phytoplankton,

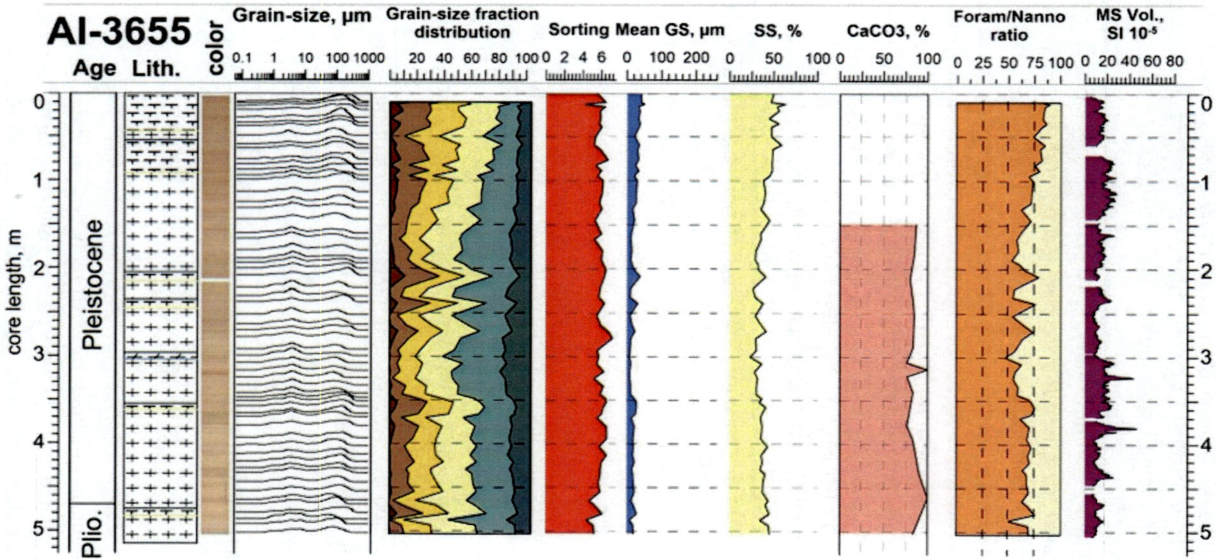

Fig. 6 Generalized age, lithostratigraphy of sediment types, color, grain-size distribution parameters (GS), $CaCO_3$ content and volume magnetic susceptibility (MS Vol.) of core AI-3655. For symbols, see Fig. 1

Plate 5 Photo of core AI-3655. [extended] means that the core section was slightly extended during storage

respectively). Unlike the opposite relationship between coarse and fine fractions in the mainly nannofossil pelagic calcareous oozes deposited without considerable influence of bottom-currents, calcareous contourites are characterized by more-or-less expressed selective reworking by bottom-currents, with the removal of nannofossils or total erosion resulting in the formation of hiatuses.

The grain-size distribution (GSD) of the high-calcareous oozes corresponds mainly to silt and sandy silt, according to the modified classical Wentworth scale (Wentworth 1922; Friedman and Sanders 1978; Blatt et al. 1980). This scale is widely used in modern international publications, including *Initial Reports of Deep-Sea Drilling Projects*. The overly wide range of the silt fraction sizes (2–63 μm) makes it challenging to use this scale on real biogenic calcareous oozes. Accordingly, we propose a subdivision of the silt fraction into clearly particulate sortable silt (*SS*) (McCave et al. 1995, 2017; McCave 2008), mainly composed of strongly dissolved fragments of foraminiferal tests (10–63 μm), and fine silt (*FS*) that contains intact nannofossils (very thin calcareous plates of coccolithophore algae) together with their fragments. Of course, the *FS* may also contain small, barely discernable foraminiferal test fragments, yet nannofossil material strongly dominates, as shown by smear-slide and SEM examination of selected samples.

The selected size fractions content (Figs. 1, 2, 3, 4, 5 and 6) along the core section shows that the silt fraction (2–63 μm) dominates throughout. The sand fraction (63–500 μm) occurs in high concentrations only in thin interlayers but is a common second order grain-size component in many core intervals from the drift summit and upper slope. Clay size fraction (<2 μm) not studied here in details forms an admixture in most studied samples. The silt fraction is subdivided into sortable silt (*SS*, 10–63 μm) and fine silt (*FS*, 2–10 μm).

The *SS* content ranges from the minimum value of 1.5% in core AI-3317 to as high as 54% measured in core AI-2436 (Table 1). Its mean values vary between 25 and 39% in the various cores and thus, in most cases, belong to sediment-forming fractions. The relatively high *SS* content in cores from the Ioffe Drift summit suggests bottom-current influence on the deposition of calcareous contourites, as shown for terrigenous ones (McCave et al. 1995, 2017; McCave 2008), because calcareous *SS* particles (foraminiferal fragments) can be transported in suspension by a bottom-current to settle selectively, depending on its velocity.

Moreover, following McCave et al. (1995, 2017), we believe that the calcareous *SS* content and its mean diameter play a crucial role in identifying the influence of relative variations in bottom-current velocity on the biogenic calcareous ooze deposition, as shown earlier for terrigenous sedimentation.

The content of *FS* varies from the very low value of 0.04% in a sand interlayer in core AI-3316 to as high as 64% in a fine-grained interlayer in core AI-3317; however, its average value for the entire set of analyzed samples (32.11%) is close to the mean *SS* content. The *FS* seems to behave like the clay-size fraction during sediment transport, together forming rather stable suspensions.

The content of the sand-size fraction (63–500 μm) is rather variable, showing high peaks on saw-shaped oscillations over a relatively low background level, contrasting between individual cores (Figs. 1, 2, 3, 4, 5 and 6) The sand-size fraction content across the entire analyzed sample set from the upper Ioffe Drift ranges from 0–99%, averaging 29.33%. The sand fraction coarser than 100 μm mainly consists of intact planktic foraminiferal tests and relatively large fragments (broken tests). This was confirmed by examination of the >0.1 mm fraction selected for foraminiferal analyses by binocular microscope (see Chap. 7). The sub-fraction of finest sand (63–100 μm) is mainly composed of foraminiferal test fragments with a minor content of small intact tests.

In all analyzed samples from five cores collected on the upper Ioffe Drift the content of the foraminiferal component size fraction (10–500 μm) ranges from a minimum value of 63% in core AI-3317 to 89% in core AI-3655, averaging 72.77% (Table 1, Figs. 1, 2, 3, 4, 5 and 6). Only the deepest core AI-3319 from the base of the ridge slope shows a lower mean value of 55% as the dissolution of the foraminiferal tests below the lysocline may have affected the foraminiferal component content. Therefore, foraminiferal oozes strongly dominate among the sediments, at least in the uppermost parallel-stratified seismic unit of the Ioffe Drift that was recovered by five cores.

The finest clay size fraction (<2 μm) contains an indefinite mixture of micritic biogenic particles (perhaps mainly crystallites) and almost all non-calcareous (terrigenous) material, including clay minerals, finely dispersed quartz and possibly also feldspars. Its content ranges from 0–35% in the analyzed samples. Average values in individual cores never rise above the thresholds (25%) crucial to sediment classification (Table 1).

The average value for mean particle size varies between the cores from 21–42 μm and is 26.61 μm in all samples from the upper drift. Only in core AI-3319 is it much less (12 μm), apparently due to the high terrigenous clay content (Table 1).

The average sorting coefficient (4.4–5.9) in all the cores indicates poor sorting, a characteristic of contourites (Brackenridge et al. 2018; Stow et al. 2019), although well-sorted thin interlayers with a sorting coefficient below 2 occur in several cores (Figs. 1, 2, 3, 4, 5 and 6). The poor sorting is possibly related to the common bimodality or polymodality of the GSD, which reflect, in turn, the various transport and deposition mechanisms suggested for the foraminiferal sand and finer fractions in all the cores. We assume that the well-preserved foraminiferal tests moved from local sources along the bottom by rolling and saltation, whereas silt and clay fractions including *SS*, possibly from remote sources, settle from near-bottom suspension. Both processes are forced by bottom-currents but apparently at different levels: at the bottom surface as foraminiferal sand; and over the bottom at various distances within suspension clouds.

3.2 End-Member Modeling of the Grain-Size Distribution

The GSDs of most sediment samples is highly complicated and polymodal (Fig. 7. There is a constant minimum interval of 10–11 μm almost in all GSDs of the cores studied. The GSDs of the finer part (<10 μm) in all cores apart from AI-3316 are typically characterized by a single pronounced maximum (the main fine mode) and by one or two subordinate modes. The main fine mode ranges between 2 and 6 μm. Core AI-3316 is characterized by two equal modes in the fine fraction, at 3 and 7 μm. In the GSDs of the coarse part, the modes are strongly variable.

To subdivide the GSDs to genetically meaningful subpopulations, end-member (EM) modeling was performed. The number of the EMs and degree of their conjunction with the given dataset were estimated by goodness-of-fit statistics. Three parameters were considered: the linear correlation between the measured GSDs and the modeled EMs; the linear correlation between the EMs; and the angular deviation between the dataset and EMs. The best-modeled EMs should have a high correlation with the data, a low correlation between EMs and a low angular deviation from the dataset (Weltje, 1997; Weltje & Prins, 2007). In our case, the best correlation is reached if three EM are chosen for all samples studied (Fig. 8).

According to the goodness-of-fit statistics, three- and four-modeled EMs best describe the GSDs in cores AI-3316, AI-3318 and AI-3317 (Fig. 8). These models

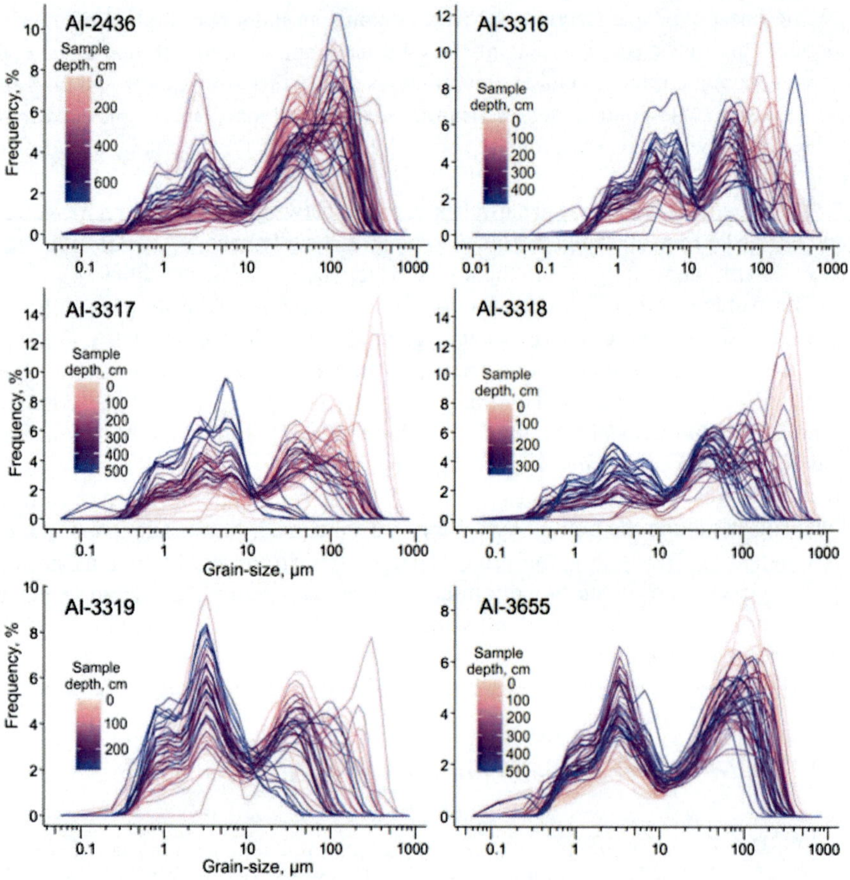

Fig. 7 The analyzed individual grain-size distributions in cores from the Ioffe Drift Note a rather stable minimum about 10–11 μm

display a low correlation between EMs and a high value of the coefficient of determination (r^2) between the data and the EMs. For three EMs, r^2 varies from 0.92 to 0.93 and for four EMss, r^2 is 0.96 in each core. The angular deviation between the modeled EMs and the data is lower for the four EMs than for the three, as expected; the average value is 11.2 for three EMs, and 8.0 for four EMs. Nonetheless, the fourth EM consists of three other EMs and the addition does not enhance the modeling, therefore it was concluded that the GSDs of AI-3316, AI-3318 and AI-3317 can be presented as a mixture of three EMs.

In cores AI-2436, AI-3655 and AI-3319, the use of two and three EMs best describes the data (Figs. 8 and 9). The coefficient of correlation between the two-modeled EMs and the dataset averages 0.86 for the three cores and higher for the three-modeled EMs, at 0.95. The mean value of the angular deviation of the EMs decreases from 14.3 for two EMs to 8.9 for three EMs. A further increase in the

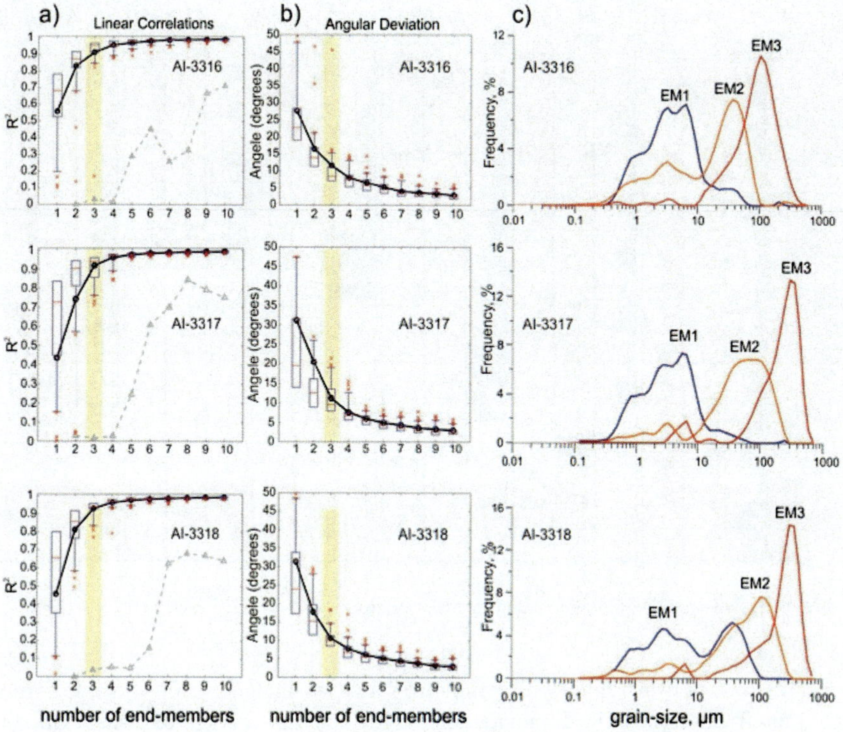

Fig. 8 End-member modeling results for the studied cores showing the goodness-of-fit statistics for the different various numbers of end-members: **a** squared linear correlations as a function of the number of EMs; **b** angular differences (in degrees) between the reconstructed EM and observed data sets as a function of the number of end-members; **c** modeled end-members, according to the four-end-member model

number of EMs results in a sharp increase in r^2 between EMs, indicating that the data have been overfitted. Consequently, the three-EMs model was chosen to describe the GSD dataset of sediment cores AI-2436, AI-3655 and AI-3319. It is worth noting that choosing the three-EMs model for all the cores helped to homogenize the results and made the interpretation easier.

The unambiguous interpretation of end-members can be challenging. EM modeling considers these three hydraulically dissimilar components to be: EM1 (stable suspension); EM2, mainly *SS* (selective deposition from suspension); and EM3, foraminiferal sand (moving along the bottom, not included into suspension, but driven by the same general contour current from above. The EM reflect a variability of GSD in the core sections that results in bimodality of the model EM curves not typical for more uniform GSD models. For example, bimodal EM1 in AI-2436 and AI-3318, or bimodal EM2 in AI-3316 and AI-3319 (Fig. 8) reflect inclusion of layers with contrasting GSD rather than subpopulation patterns.

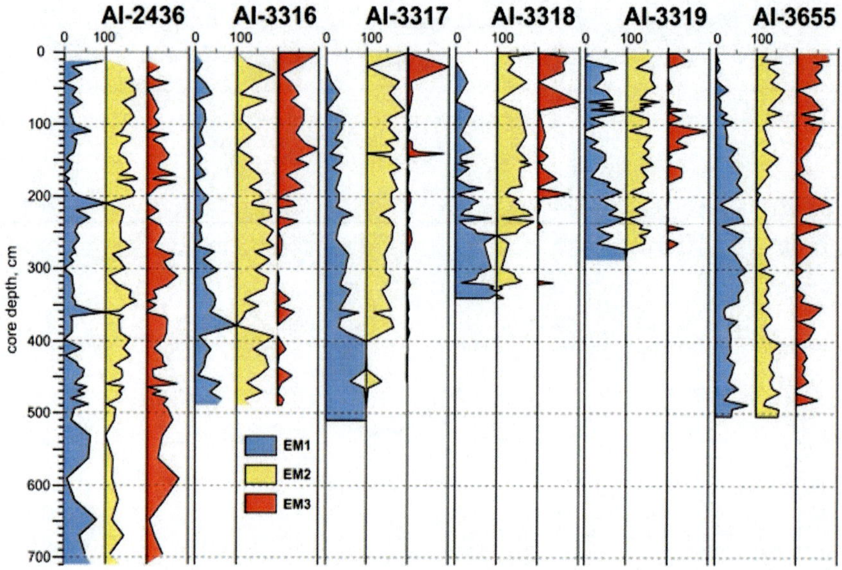

Fig. 9 Down-core distributions of end-member values

The EM2 (Fig. 8), along with common *SS* fraction, includes a considerable portion of sand fraction (up to its domination in the EM2 model of AI-3318). Association of fine-grained sand with dominating *SS* in the EM2 really exists, as shown by many individual GSD analyzes. This means that selective settling of foraminiferal sand (likely lightweight intact tests) together with *SS*-size foraminiferal fragments, from the near-bottom suspension correctly interprets the EM2 model.

As shown by Fig. 9, EM1 content in different cores generally follows the fine grained (<10 μm) content (see Figs. 1, 2, 3, 4 and 5). High values occur only in lower nannofossil-rich layers of cores AI-3317 and AI-3318, less apparently in lower parts of cores AI-3316 and AI-2436, whereas EM1 content varies at low levels in upper layers of all these cores. Sharp oscillations at a moderately high level occur in the deep-water core AI-3319. Therefore, a restricted contribution of EM1 at least to the Quaternary sections of the Ioffe Drift summit and upper slope suggests a limited deposition from suspension clouds in relatively still bottom water conditions. More favorable environments for EM1 accumulation existed over the drift in the Upper Pliocene. Short-term still-water episodes between high-energy stages led to distinct thin fine-grained interlayers of EM1 deposition at the AI-2436, AI-3317, AI-3318 sites, and also at the base of the drift (AI-3319). The nature of these short-term still-water events needs a further study.

The high EM2 content in many cases prevailing EM1 content, characterizes Quaternary high-calcareous sediments from the Ioffe Drift summit and upper slope (Fig. 9, cores AI-3317, AI-3318, AI-2436). The EM2 content commonly prevails over that of EM1 also in core AI-3316, but at a lower absolute level. Relatively lower content of EM2 is measured in cores AI-3655 and AI-3319. We suggest that the amount of EM2 mainly depends on a delivery of *SS* and fine-grained sand size foraminiferal fragments to the drift from deeper areas below the lysocline. Of course, their transport in the turbulent suspension and selective deposition from the suspension need appropriate bottom water energy.

EM3 content is extremely variable, from 0 to 80% or more (Fig. 9) showing sharp peaks at foraminiferal sand interlayers. More or less continuous EM3 contribution, although variable, is documented in cores AI-2436 and AI-3655, whereas high values occur only in the uppermost layers of cores AI-3316, AI-3317, AI-3318. Small rare EM3 peaks occur in core AI-3319. We assume that the EM3 distribution is mainly related to high bottom-currents velocities, which remove finer grained material from the foraminiferal sand, rather than to any material supply.

3.3 Correlation of SS Content in Fractions <63 μm and Mean SS Particles Grain Size

According to McCave et al. 2017, this correlation indicates an influence of bottom-currents on the terrigenous sediment deposition thus distinguishing contourites from pelagic (hemipelagic) sediments. Our approach illustrated by Fig. 10 first shows that the criterion is also valid for calcareous contourites. Indeed, correlation coefficients of calcareous (foraminiferal) *SS* content in <63 μm and mean *SS* particle size in all analyzed six cores, ranging from 0.686 in core AI-2436 (drift summit) to 0.895 in core AI-3319 (series of wave-like depositional features to the WSW from the Ioffe Drift) are high enough to support their prevailing contourite origin. Only in core AI-3317 we removed a point plotted far from the general regression line to reach the coefficient of 0.828. The rather low correlation coefficient in core AI-2436 possibly expresses preliminarily sorted *SS* contribution to its budget in sediments highly enriched in calcareous *SS*. Note, core AI-3655 from almost the same site has the coefficient 0.783 with much lower maximum *SS* content.

The maximum mean *SS* particle size does not exceed 47 μm (core AI-3318) and the lowest maximum value, 35 μm is measured in core AI-3319 (Fig. 10). These narrow limits likely characterize the prevailing size of dissolved foraminiferal fragments delivered from an area below the lysocline. The percentages of *SS* in <63 μm reach up to almost 100% in core AI-3316 possibly owing to the highest bottom-currents velocity but reached high values also at sites of cores AI-3318 (maximum about 89% *SS*) and AI-2436 (maximum about 88% *SS*). However, the percentages of *SS* were also controlled by its content and grain size in the suspension clouds washing the Ioffe Drift sites.

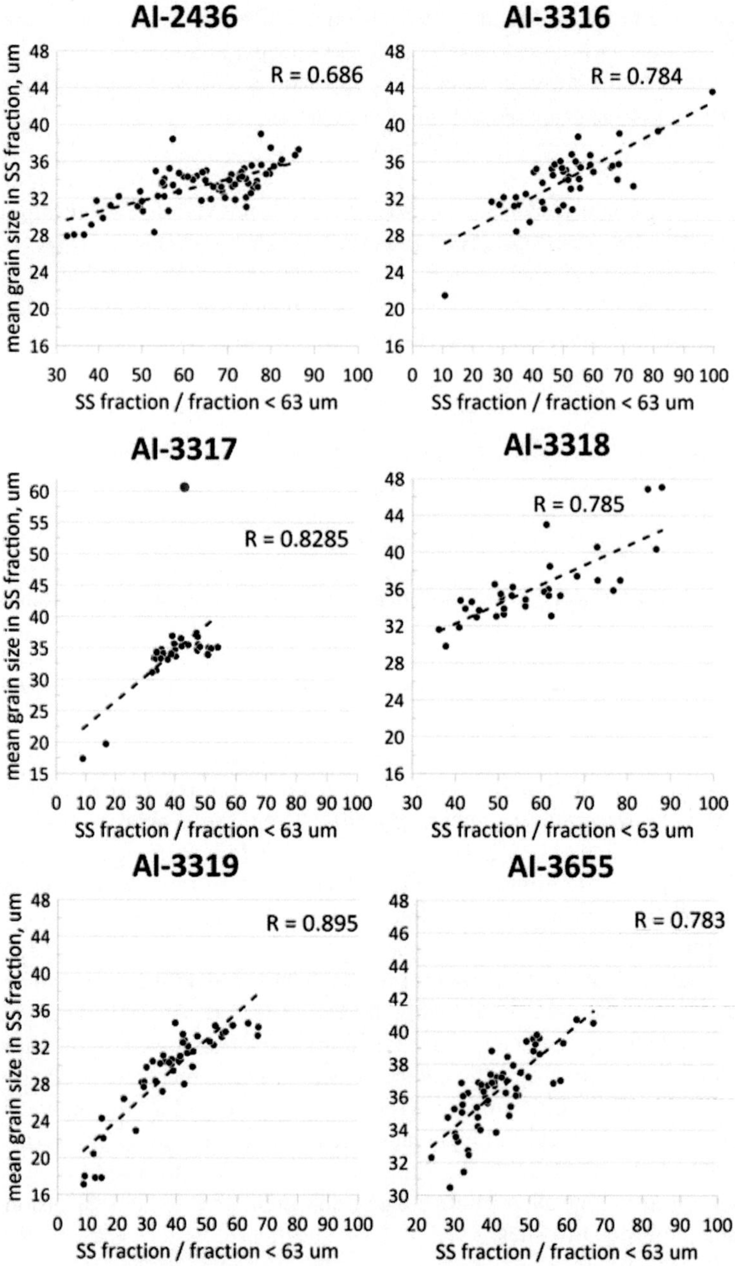

Fig. 10 Diagrams showing correlation between mean size particles of sortable silt and amount percentage of sortable silt in <63 μm fraction. The red dot marks the sample excluded from the correlation coefficient calculation (the explanation is given in text)

Table 2 Minimum, maximum and mean content of $CaCO_3$ and total organic carbon in the studied cores

Core/parameter	AI-2436		AI-3316		AI-3317		AI-3318		AI-3319	AI-3655
	$CaCO_3$	TOC	$CaCO_3$	TOC	$CaCO_3$	TOC	$CaCO_3$	TOC	$CaCO_3$	$CaCO_3$
Min, %	73.51	0.11	79.89	0.07	82.93	0.06	90.88	0.02	19.38	76.61
Max, %	90.39	0.55	100.01	0.33	98.88	0.26	99.61	0.14	62.39	100.00
Mean, %	82.68	0.24	92.39	0.18	91.44	0.10	95.87	0.05	42.25	84.96

4 Geochemistry ($CaCO_3$, TOC, XRF)

All the studied cores on the Ioffe Drift are characterized by a high $CaCO_3$ content varying from 73 to 100%. Calcium carbonate content in core AI-3319 changes between 19% and 62.5% (Table 2).

The TOC content is generally low (0.14% on average). Only in the upper 500 cm of core AI-2436 it reaches 0.55% (Table 2), that is not low for calcareous pelagic sediments commonly containing much less organic matter. In this part of core, the relatively increased TOC content demonstrates correlation with clay amount. This correlation was not revealed in the rest of cores (Figs. 2, 3 and 4). The difference in TOC content can be explained by the higher sedimentation rates during the deposition of the upper 500 cm of core AI-2436. The almost similar $CaCO_3$ and TOC contents were measured in surface sediments from the Rio Grande Rise, namely 70–90% of $CaCO_3$ and 0–0.5% of TOC (Frenz et al. 2003).

The Ca/Al and Ca/Fe ratios reflect the balance of biogenic calcareous and lithogenic components of the sediments (Blanchet et al. 2009). The normalized values of Si and Al demonstrate a good correlation with the MS record and mark the intervals which are enriched with terrigenous material (Fig. 11).

The S/Cl ratio associated with pyrite or organic carbon (Thomson et al. 2006) does not correlate to the TOC content. This proxy probably does not work in calcareous oozes within the strongly oxic sedimentation and early diagenesis environments characteristic for the pelagic realm. The low amplitude of the variations in the Mn/Fe ratio (Fig. 11) indicates oxic conditions in the sediments of all cores (Marsh et al. 2007).

5 Magnetic Susceptibility and Color Reflectance

MS is the degree of magnetization in response to an applied magnetic field. It is one of important sediment properties to be routinely logged during core analysis and reported in dimensionless units. Generally, it reflects thc Fc and Ti mineral concentrations, with an emphasis on ferromagnetic phases (Kent 1982; Doh et al. 1988; Rothwell and Croudace 2015).

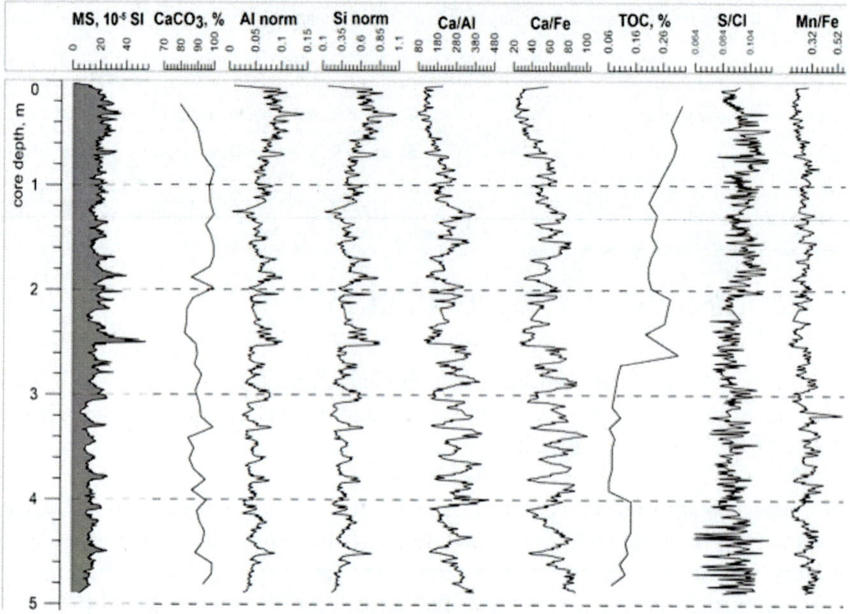

Fig. 11 Distribution of major geochemical ratios along the core AI-3316. Norm.–normalized by the total scatter)

The volume MS value of the drift's sediments varies from 5.35 to 36.11 SI 10^{-5} (median 22.087). In the deep-water core AI-3319 it changes from 5.02 to 73.872 SI 10^{-5} (median 31.29). This difference is explained by the greater amount of terrigenous material in core AI-3319. The MS records demonstrate generally similar variation patterns. The visual correlation of the available records was based on four to eight tie-points, marked down-core from Tie-point t8–t1 (Fig. 12).

According to the MS data, the cores can be subdivided into three main parts. The upper corresponds to the interval from the core top to Tie-point t8 and is characterized by relatively gradual changes in MS. The thickness of this part varies from 1.13 to 2.36 m (apart from in cores AI-2436 and AI-3317, where this pattern is barely discernable). The upper part is underlain by a thin transition zone between Tie-points t8 and t5, characterized by two clearly recognizable peaks. The middle part (between Tie-points t5 and t3) demonstrates a set of up to seven closely spaced high-amplitude peaks. This part is up to 0.75 m thick. Most of the peaks show a well-defined asymmetry with an abrupt change at the bottom, followed by a gradual upward change. The lower part is characterized by a series of high-amplitude peaks with a spacing interval of 16–38 cm (twice that in the middle part). Some peaks in the lower part demonstrate the aforementioned type of asymmetry. We consider these peaks as the signatures of erosional hiatuses (Ivanova et al. 2020; see also Chap. 9). The asymmetric peaks correspond to the darker layers, enriched in terrigenous material.

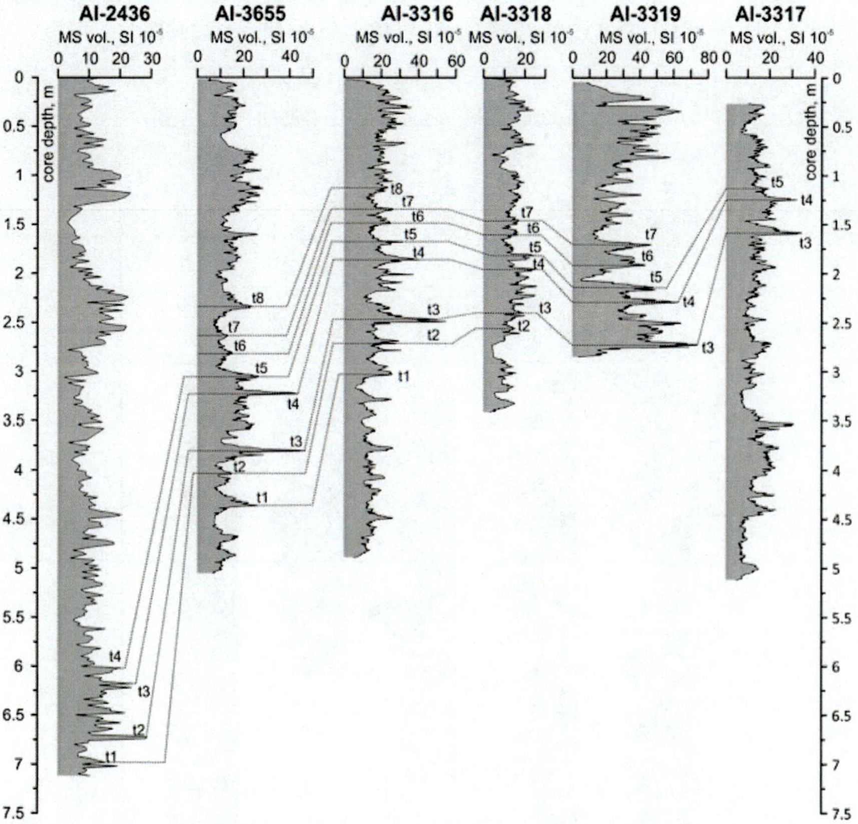

Fig. 12 Magnetic susceptibility records of the studied cores with correlation. Tie-points are marked shown as t1–t8

In cores AI-3316, AI-3318 and AI-3655, the L*-parameter records generally demonstrate a down-core lightening of sediments. These variations correlate to a slight decrease in foraminiferal abundance. Cores AI-3317 and AI-3319 are characterized by a gradual increase in L* value in the upper part and a decrease in the lower (Fig. 13). The sediments of core AI-3319 are of the darkest color, while the other cores do not differ significantly in lightness. The most distinct color differences between cores were determined on a projection of L*a*b* space on the L*b* plane (Fig. 14). The L* records show anti-phase correlation to MS and a sufficiently robust positive correlation to the Ca/Al ratio. Significant changes in the L*-value mark visible changes in sediment composition and color within the core sections, as well as the lower and upper boundaries of the intervals, with a set of high-amplitude MS peaks in the middle part of the cores, as described above. Thus, in several cases changes in L*-value also indicate erosional hiatuses (Ivanova et al. 2020; see also Chap. 9). The a*-parameter, reflecting the degree of red hue expression, correlates well to Fe content and MS (Fig. 15).

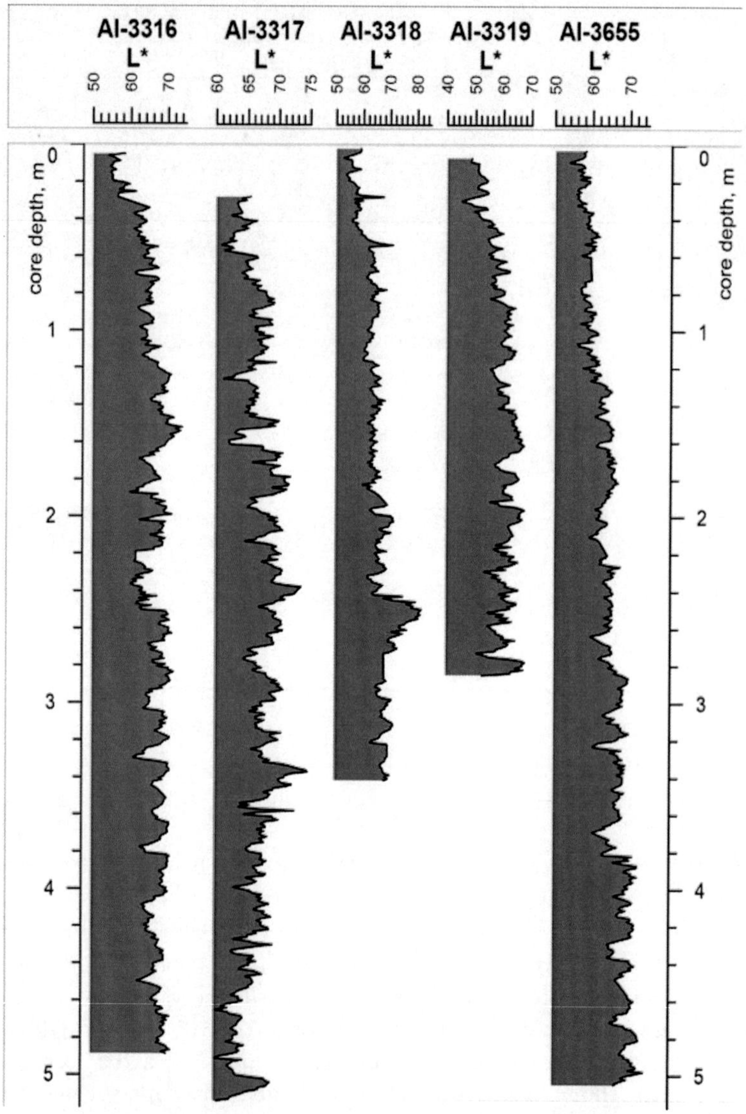

Fig. 13 Down-core variations of the L*-parameter (color reflectance)

6 Mineralogy (XRD Data)

The mineral composition of the Ioffe Drift deposits was studied in three cores: AI-2436; AI-3316; and AI-3318. The bulk mineral composition and minerals of the clay fraction were analyzed in 4 and 17 samples (intervals) from core AI-2436, respectively. Dark-colored interlayers were studied in cores AI-3316 and AI-3318.

Fig. 14 The results of spectrophotometric measurements in the CIE L*a*b* color space presented as projections on: **a** a*b* plane; **b** L*b* plane; and **c** L*a* plane. Every colored dot marks the result of one measurement

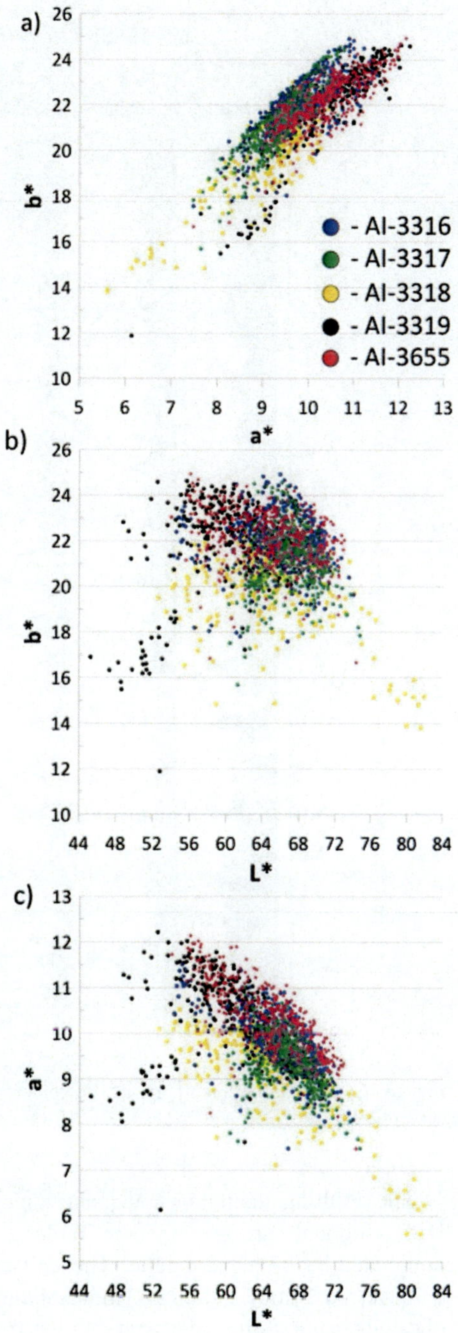

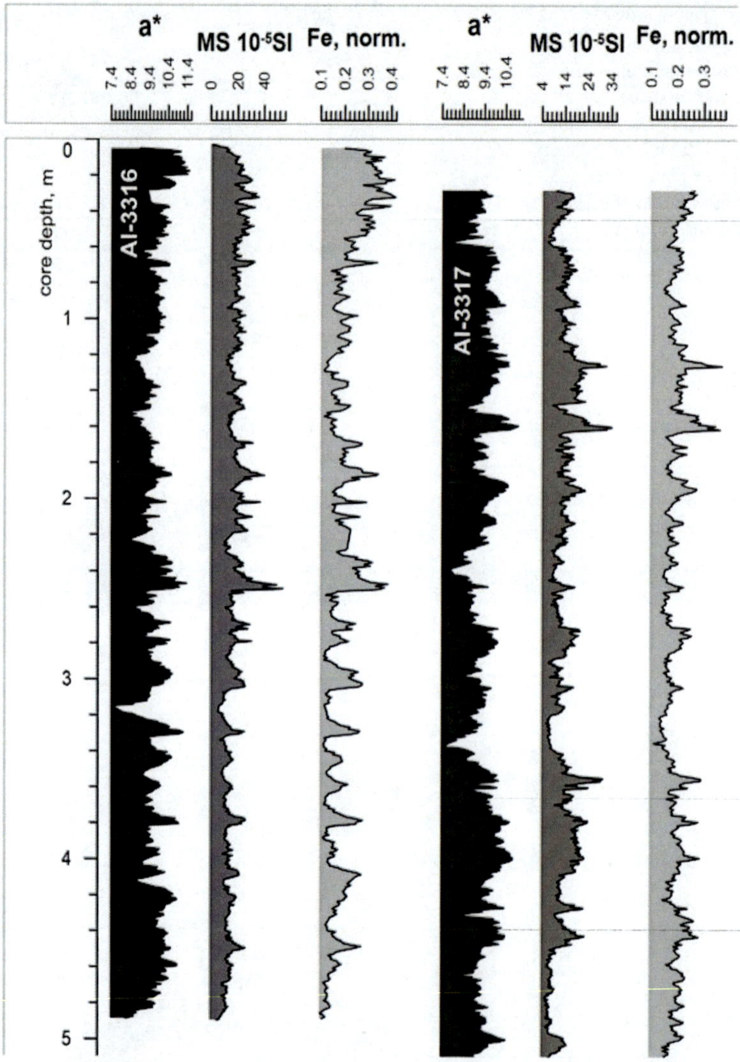

Fig. 15 Correlation between color a*, Fe content (normalized by the total scatter) and magnetic susceptibility

The resulting qualitative and quantitative mineralogical characteristics of the Ioffe Drift sediments are presented in Tables 3, 4 and 5. The mineral composition is dominated by biogenic calcite. The unit cell parameters of calcite (a = 4.989 ± 0.002 Å, c = 17.062 ± 0.002 Å) indicate minimal isomorphic impurities. The content of calcite in the studied sediment samples from core AI-2436 varies in a narrow range of 88–91%. The sum of detrital minerals (quartz, andesine, amphibole), clay minerals

Table 3 Mineral composition of sedimentary material from the Ioffe Drift (core AI-2436)

Core depth, cm	Calcite, %	Quartz, %	Andesine, %	Smectite, %	Illite, %	Kaolinite, %	Chlorite, %	Trace minerals
12–14	88	3	2	<1	2	2	1	gypsum,
322–324	91	1	1	<1	1	2	1	amphibole, zeolite, gypsum
532–534	91	1	1	<1	2	2	1	zeolite
710–712	91	1	1	<1	2	2	1	amphibole

Table 4 Clay mineral composition of sediments from the Ioffe Drift sediments (core AI-2436)

Core depth, cm	Smectite, %	Illite, %	Kaolinite, %	Chlorite, %
12–14	7	69	17	6
50–52	11	60	22	6
100–102	38	43	13	6
140–142	5	68	20	6
180–182	10	64	19	6
200–202	8	66	19	6
230–232	8	61	25	5
278–280	7	59	28	6
300–302	14	62	21	2
322–324	10	61	24	4
360–362	6	66	25	3
400–402	10	62	22	6
430–432	11	63	22	4
470–472	12	63	21	3
494–496	8	55	32	5
532–534	16	60	19	4
710–712	8	67	21	2

(smectite, illite, kaolinite, chlorite), as well as likely authigenic gypsum and zeolite does not exceed 12%.

The content of dioctahedral alkaline-earth smectite varies from 5 to 38%, trioctahedral Mg-Fe- chlorite from 2 to 6%, illite from 43 to 69% and kaolinite from 13 to 32%. Thus, the illite-dominating clay mineral assemblage with highly variable second-order smectite and kaolinite characterizes the analyzed samples from the core AI-2436.

The relationship between clay minerals (with a prevalence of illite) in the Ioffe Drift sediments is similar to sediments from the Rio Grande Rise (DSDP 516, Zimmerman 1983). It differs from deposits covering the southern and central parts

Table 5 Mineral composition (%) of sediment samples from the Ioffe Drift (cores AI-3316 and AI-3318). Note: XR-A–X-ray amorphous phase

Sample	Calcite	Quartz	Andesine	K-Fsp	Amphibole	Fe–Mn nodules	Smectite	Illite	Kaolinite	Chlorite	XR-A phase
AI-3318											
71–72	96	1	1	n.d	n.d	n.d	n.d	1	1	n.d	11
196–197	88	2	3	n.d	1	1	traces	3	1	1	17
196–197 dark grains	Birnessite and 10Å minerals of the asbolan–buserite group										n.d
232–233	94	1	1	n.d	n.d	n.d	1	1	1	1	12
250–251	96	1	n.d	n.d	n.d	1	n.d	1	1	n.d	14
250–251 dark grains	Buserite 1, asbolan										48
AI-3316											
8–9	91	3	2	n.d	n.d	n.d	n.d	2	1	1	28
70–71	92	2	2	n.d	1	n.d	traces	1	1	1	11
154–155	95	1	1	n.d	n.d	1	n.d	1	1	n.d	12
154-155dark grains	Talc, quartz, 7Å phase, possibly birnessite										80
245–246	92	2	1	1	1	n.d	traces	1	1	1	13
255–256	96	1	1	n.d	n.d	n.d	n.d	1	1	traces	12
298–299	95	2	1	n.d	n.d	n.d	traces	1	1	n.d	15

**n.d.–not detected*

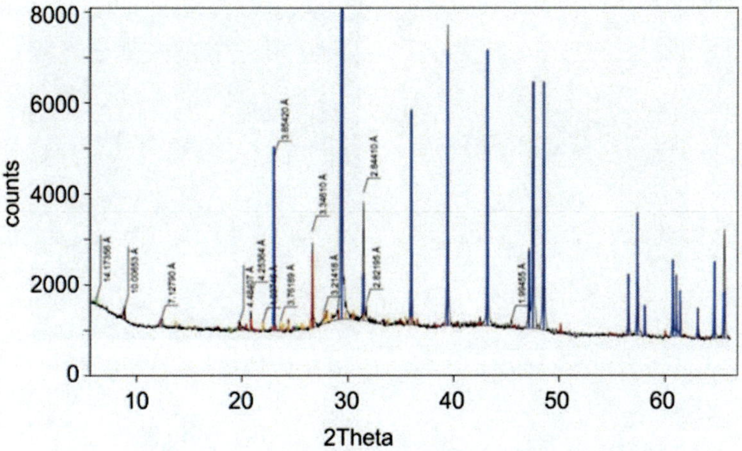

Fig. 16 Diffraction pattern of a sediment sample from core AI-3316 (8–9 cm)

of the Brazil Basin (DSDP 515, Zimmerman 1983), where minerals of the smectite group dominate. According to Windom (1976), the increased amount of illite (>50%) in the studied part of the South Atlantic marks the impact of the Rio de la Plata input. However, illite can be delivered to the study area as aeolian material. Patagonia is a potential source area of the illite-enriched dust, derived from loess deposits (Zárate and Blasi 1993; Gingele et al. 1999).

Figure 16 shows a diffraction pattern of a typical sediment sample from the study area. The amount of calcite in bulk samples from cores AI-3318 and AI-3316 varies from 88 to 96% and from 92 to 96%, respectively. The total content of terrigenous clastic minerals, including quartz, microcline, albite, Ca-plagioclase, amphibole, varies from 2 to 7%. The total content of minerals of the clay group: illite, kaolinite, chlorite, smectite and mixed-layered varieties do not exceed 5%.

The visually distinguishable dark-colored inclusions picked from the sediments were photographed and, if there was a sufficient amount of grains, studied by X-ray phase analysis. Plate 6a, b shows an image of dark-colored smears and localized grains from core AI-3316 at the interval of 154–155 cm, and Plate 6c shows dark-colored grains from the sediments of the same core at the interval 245–246 cm. Plate 6d shows the micro-nodules retrieved from the interval 196–197 cm of core AI-3318, while images of micro-nodules picked from the sediments of the same core at the interval 250–251 cm are presented in Plate 7.

According to the X-ray phase analysis data, dark-colored smears in the sediments from the 154–155 cm interval (core AI-3316) comprise X-ray-amorphous material. The grains associated with this substance are aggregates consisting of quartz, muscovite-type mica and talc. A weak 7 Å peak may indicate the presence of birnessite. The dark X-ray amorphous material coating foraminiferal tests in this interval is similar to the material of dark-colored grains from the interval 245–246 cm of the same core. The micro-nodules extracted from the interval 196–197 cm of core

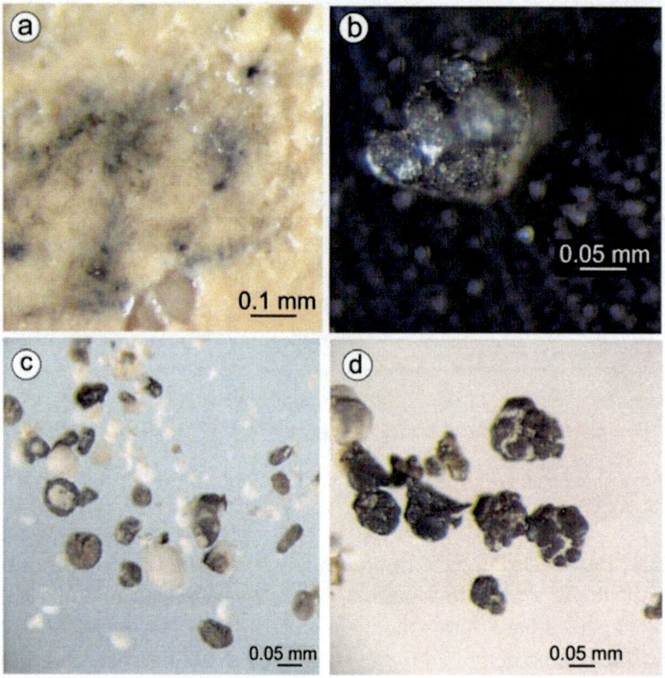

Plate 6 Image of **a** dark-colored smears and **b** grains localized in them from core AI-3316 (154–155 cm); **c** dark-colored grains taken from the interval 245–246 cm of core AI-3316; **d** image of micronodules from core AI-3318 (196–197 cm)

AI-3318 include birnessite and 10 Å minerals of the asbolan-buserite group. X-ray analysis of Fe–Mn micro-nodules from core AI-3318 (interval 250–251 cm) revealed buserite-1 with sufficient confidence since, when the sample is heated (105 °C), the 10 Å transforms entirely into the birnessite phase (Fig. 17, II, III; peak 9.99 Å refers to mica) The content of such material (about 1%) is below the detection limit for a bulk sample (Fig. 17, I).

7 Conclusions

The lithology of the calcareous Ioffe Drift differs from its terrigenous analogs of similar GSDs in terms of its composition (entirely low-magnesium calcite) and the initial biomorphic shape (planktic foraminiferal tests, calcareous nannofossils) of the sediment-forming particles. Unlike the terrigenous material, the initial biogenic particles were produced mainly in surface waters. Their size and shape may be changed during sedimentation: they can be broken mechanically or partially dissolved to produce fragments of varying size and shape. The subsequent behavior of both the

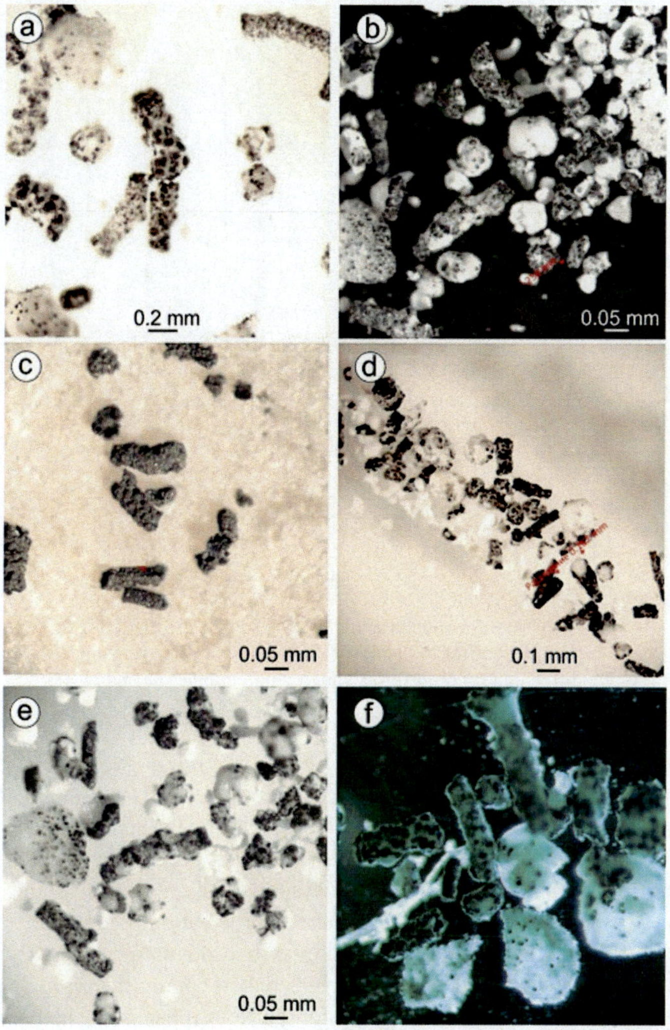

Plate 7 Images of micronodules from core AI-3318 (250–251 cm); **a** and **e** micronodules in intergrowth with biogenic calcareous material, detritus, and foraminifera tests on a white background; **b** micronodules in intergrowth with biogenic carbonate material, detritus, and foraminifera tests on a black background; **c** micronodules after treatment with hydrochloric acid; **d** general view; **f** micronodules in intergrowth with biogenic calcareous material, detritus, and foraminifera tests by transmission of light

intact biomorphic particles and their fragments resembles that of terrigenous material, and the sedimentological regularities of the contourite theory are equally valid for the calcareous drifts.

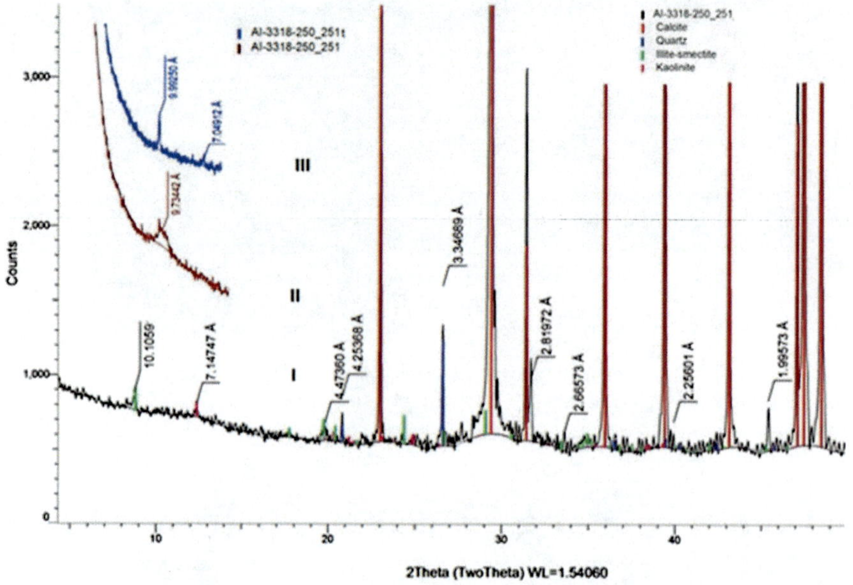

Fig. 17 Diffraction patterns of sediment samples from core AI-3318 (250–251 cm): I–diffractogram of the bulk sample; II–fragment of the diffractogram of ferromanganese micronodules; III–fragment of the diffractogram of ferromanganese micronodules heated at 105 °C

Two main processes strongly affect the Ioffe Drift lithology, notably mechanical sedimentation under the influence of bottom-currents and dissolution by the aggressive bottom waters of Antarctic origin. The dissolution influence is significant only below the foraminiferal lysocline in core AI-3319. The other five cores were retrieved from the drift summit and upper slope well above the regional lysocline, and their lithology depends entirely on the mechanical sedimentation of high-calcareous ooze mainly controlled by bottom-currents and other hydrodynamic processes.

The multi-proxy study of these cores revealed their contourite origin. In particular, GSD provided the main evidence of the contourite origin for all the cores studied, including AI-3319, where the contourite process is associated with biogenic calcium carbonate dissolution. Despite considerable facies variation between cores, including a sharp facies difference in cores AI-2436 and AI-3655 taken from practically the same site, common features in terms of GSD are apparent and were documented. Silt fractions dominate, with a considerable contribution of *SS*. The relationship between *SS* concentration and mean size suggests the influence of bottom-current velocities on the deposition of the contourite.

The characteristic poor sorting commonly noted in contourites is related in our case to bimodal or polymodal individual GSDs, which are created by contrasting transport and deposition mechanisms for the main sediment components, namely foraminiferal sand, fine foraminiferal tests fragments, mostly of *SS* size, and nannofossils together with clay-size particles. EM modeling considers these three hydraulically dissimilar components as: EM1 (stable suspension); EM2, mainly *SS* (selective deposition from suspension); and EM3, foraminiferal sand (moving along the bottom, not included in suspension).

MS data served as the basis for a reliable correlation of the studied cores. According to the MS distribution patterns correlated to color reflectance and XRF data, most of the core sections display three parts that probably correspond to differing sedimentation modes. The lower part (below Tie-point t3) demonstrates the pulsating nature of the terrigenous material supply to the drift area (and/or carbonate dissolution) with episodic erosion events. The middle part (Tie-points t3–t5) shows a series of more frequent drastic erosion cycles followed by a high (but rapidly decreasing) supply of terrigenous material (and/or carbonate dissolution). The upper part shows a gradual change in the terrigenous material supply. It is suggested that the erosion and supply of terrigenous material was mainly controlled by bottom-currents.

The main mineral in the biogenic sedimentary material of the Ioffe Drift is calcite, with a minimal amount of isomorphic impurities. The content of other minerals, notably clay and detrital terrigenous minerals does not exceed 12%. Among the dark-colored grains extracted from the sediments, intergrowths of detrital minerals, foraminiferal tests with a dark X-ray amorphous cover and ferromanganese micronodules were detected.

References

Balsam WL, Damuth JE, Schneider RR (1997) Comparison of shipboard versus shore-based spectral data from Amazon Fan cores: implications for interpreting sediment composition. In: Proceedings of the Ocean Drilling Program, 155 scientific results. Ocean Drilling Program, pp 531–538

Berntsson A, Rosqvist GC, Velle G (2014) Late-Holocene temperature and precipitation changes in Vindelfjällen, mid-western Swedish Lapland, inferred from chironomid and geochemical data. Holocene 24:78–92. https://doi.org/10.1177/0959683613512167

Blanchet CL, Thouveny N, Vidal L (2009) Formation and preservation of greigite (Fe3 S4) in sediments from the Santa Barbara Basin: Implications for paleoenvironmental changes during the past 35 ka. Paleoceanography 24: n/a-n/a. https://doi.org/10.1029/2008PA001719

Blatt H, Middleton GV, Murray RC (1980) Origin of sedimentary rocks. Prentice-Hall

Brackenridge RE, Stow DAV, Hernández-Molina FJ et al (2018) Textural characteristics and facies of sand-rich contourite depositional systems. Sedimentology. https://doi.org/10.1111/sed.12463

Brown G (ed) (1961) The X-ray identification and crystal structures of clay minerals. Mineralogical Society, Clay Minerals Group

Doh S-J, King JW, Leinen M (1988) A rock-magnetic study of giant piston core LL44-GPC3 from the central North Pacific and its paleoceanographic implications. Paleoceanography 3:89–111. https://doi.org/10.1029/PA003i001p00089

Frenz M, Höppner R, Stuut J-BW et al (2003) Surface sediment bulk geochemistry and grain-size composition related to the oceanic circulation along the South American continental margin in the Southwest Atlantic. The South Atlantic in the late quaternary. Springer, Berlin Heidelberg , pp 347–373

Friedman GM, Sanders FE (1978) Principles of sedimentology. Wiley, New York

Gingele FX, Schmieder F, von Dobeneck T et al (1999) Terrigenous Flux flux in the Rio Grande Rise area during the past past 1500 ka: evidence of deepwater advection or rapid response to continental rainfall patterns? Paleoceanography 14:84–95. https://doi.org/10.1029/1998PA900012

Giosan L, Flood RD, Grützner J, Mudie P (2002) Paleoceanographic significance of sediment color on western North Atlantic drifts: II. Late Pliocene-Pleistocene sedimentation. Mar Geol 189:43–61. https://doi.org/10.1016/S0025-3227(02)00322-5

Guyard H, Chapron E, St-Onge G et al (2007) High-altitude varve records of abrupt environmental changes and mining activity over the last 4000 years in the western French Alps (Lake Bramant, Grandes Rousses Massif). Quat Sci Rev 26:2644–2660. https://doi.org/10.1016/j.quascirev.2007.07.007

Hamann Y, Ehrmann W, Schmiedl G et al (2008) Sedimentation processes in the Eastern Mediterranean Sea during the Late Glacial and Holocene revealed by end-member modelling of the terrigenous fraction in marine sediments. Mar Geol 248:97–114. https://doi.org/10.1016/j.margeo.2007.10.009

IJmker J, Stauch G, Dietze E et al (2012) Characterisation of transport processes and sedimentary deposits by statistical end-member mixing analysis of terrestrial sediments in the Donggi Cona lake catchment, NE Tibetan Plateau. Sediment Geol 281:166–179. https://doi.org/10.1016/j.sedgeo.2012.09.006

Ivanova E, Murdmaa I, Borisov D, Dmitrenko O, Levchenko O, Emelyanov E (2016) Late Pliocene-Pleistocene stratigraphy and history of formation of the Ioffe calcareous contourite drift, Western South Atlantic. Mar Geol 372:17–30. https://doi.org/10.1016/j.margeo.2015.12.002

Ivanova E, Borisov D, Dmitrenko O, Murdmaa I (2020) Hiatuses in the late Pliocene–Pleistocene stratigraphy of the Ioffe calcareous contourite drift, western South Atlantic. Mar Pet Geol 111. https://doi.org/10.1016/j.marpetgeo.2019.08.031

Just J, Heslop D, Von Dobeneck T et al (2012) Multiproxy characterization and budgeting of terrigenous end-members at the NW African continental margin. Geochem Geophys Geosyst 13:1–18. https://doi.org/10.1029/2012GC004148

Kent DV (1982) Apparent correlation of palaeomagnetic intensity and climatic records in deep-sea sediments. Nature 299:538–539. https://doi.org/10.1038/299538a0

Kylander ME, Ampel L, Wohlfarth B, Veres D (2011) High-resolution X-ray fluorescence core scanning analysis of Les Echets (France) sedimentary sequence: New insights from chemical proxies. J Quat Sci 26:109–117. https://doi.org/10.1002/jqs.1438

Marsh R, Mills RA, Green DRH et al (2007) Controls on sediment geochemistry in the Crozet region. Deep Sea Res Part II Top Stud Oceanogr 54:2260–2274. https://doi.org/10.1016/j.dsr2.2007.06.004

Martín-Chivelet J, Fregenal-Martínez MA, Chacón B (2008) Traction structures in contourites. In: Rebesco M, Camerlenghi A (eds) Contourites. developments in sedimentology, vol 60, pp 157–182

McCave IN (2008) Size sorting during transport and deposition of fine sediments: sortable silt and flow speed. Dev Sedimentol 60:121–142

McCave IN, Manighetti B, Robinson SG (1995) Sortable silt and fine sediment size/composition slicing: parameters for palaeocurrent speed and palaeoceanography. Paleoceanography 10:593–610. https://doi.org/10.1029/94PA03039

McCave IN, Thornalley DJR, Hall IR (2017) Relation of sortable silt grain-size to deep-sea current speeds: calibration of the 'mud current meter.' Deep Sea Res Part I Oceanogr Res Pap 127:1–12. https://doi.org/10.1016/j.dsr.2017.07.003

Moore DM, Reynolds RC (1997) X-ray diffraction and the identification and analysis of clay minerals, 2nd edn. Oxford University Press, New York

Paterson GA, Heslop D (2015) New methods for unmixing sediment grain size data. Geochem Geophys Geosyst 16:4494–4506. https://doi.org/10.1002/2015GC006070

Paterson GA, Heslop D (2017) Software for grain size unmixing and analysis

Prins MA, Bouwer LM, Beets CJ et al (2002) Ocean circulation and iceberg discharge in the glacial North Atlantic: Inferences from unmixing of sediment size distributions. Geology 30:555–558. https://doi.org/10.1130/0091-7613(2002)030%3c0555:OCAIDI%3e2.0.CO;2

Rebesco M, Hernández-Molina FJ, Van Rooij D, Wåhlin A (2014) Contourites and associated sediments controlled by deep-water circulation processes: state-of-the-art and future considerations. Mar Geol 352:111–154. https://doi.org/10.1016/j.margeo.2014.03.011

Rothwell RG, Croudace IW (2015) Twenty years of XRF core scanning marine sediments: what do geochemical proxies tell us? In: Croudace I, Rothwell R (eds) Micro-XRF studies of sediment cores. Developments in paleoenvironmental research, vol 17. Springer, Dordrecht, pp 25–102

Saito Y (1996) Grain-size and sediment-color variations of Pleistocene slope sediments off New Jersey. In: Proceedings of the ocean drilling program, 150 scientific results. Ocean Drilling Program.

Shala S, Helmens KF, Jansson KN et al (2014) Palaeoenvironmental record of glacial lake evolution during the early Holocene at Sokli, NE Finland. Boreas 43:362–376. https://doi.org/10.1111/bor.12043

Shepard FP (1954) Nomenclature based on sand-silt-clay ratios. SEPM J Sediment Res 24:151–158. https://doi.org/10.1306/D4269774-2B26-11D7-8648000102C1865D

Smetannikova OG (1988) Guidelines for X-ray study of manganese oxides and hydroxides. Leningrad State University (in Russian), Leningrad

Stow DAV, Faugères J-C (2008) Contourite facies and the facies model. In: Rebesco M, Camerlenghi A (eds) Contourites. Developments in sedimentology, chap 13, 60. Elsevier, Amsterdam, pp 223–256

Stow DAV, Smillie Z, Pan J, Esentia I (2019) Deep-sea contourites: sediments and cycles. In: Encyclopedia of ocean sciences. Elsevier, pp 111–120

Thomson J, Croudace IW, Rothwell RG (2006) A geochemical application of the ITRAX scanner to a sediment core containing eastern Mediterranean sapropel units. Geol Soc London Spec Publ 267:65–77. https://doi.org/10.1144/GSL.SP.2006.267.01.05

Weltje GJ (1997) End-member modeling of compositional data: numerical-statistical algorithms for solving the explicit mixing problem. Math Geol 29:503–549. https://doi.org/10.1007/BF02775085

Weltje GJ, Prins MA (2007) Genetically meaningful decomposition of grain-size distributions. Sediment Geol 202:409–424. https://doi.org/10.1016/j.sedgeo.2007.03.007

Weltje GJ, Prins MA (2003) Muddled or mixed? Inferring palaeoclimate from size distributions of deep-sea clastics. Sediment Geol 162:39–62. https://doi.org/10.1016/S0037-0738(03)00235-5

Wentworth WC (1922) A scale of grade and class terms for clastic sediments. J Geol 30:377–392

Wetzel A, Werner F, Stow DAV (2008) Chapter 11 Bioturbation and biogenic sedimentary structures. In: Rebesco M, Camerlenghi A (eds) Contourites. Developments in sedimentology, 60, Contourites, chap 11. Elsevier, pp 183–202

Windom H.L. (1976) Lithogenous material in marine sediments. In: Riley JP, Chester R (eds) Chemical oceanography, 2nd edn. Academic Press, Inc., pp 103–135

Zárate M, Blasi A (1993) Late Pleistocene-Holocene eolian deposits of the southern Buenos Aires province, Argentina: a preliminary model. Quat Int 17:15–20. https://doi.org/10.1016/1040-6182(93)90075-Q

Zimmerman HB (1983) Clay mineral stratigraphy of the Rio Grande Rise and Southern Brazil Basin, Western South Atlantic Ocean. In: Initial reports of the deep-sea drilling project, 72. U.S. Government Printing Office

Micropaleontology and Biostratigraphy

Elena Ivanova and Olga Dmitrenko

Abstract To stratify six sediment cores, we applied the calcareous plankton biostratigraphy of the upper sediment cover based on the foraminiferal and nannofossil zonations suggested in Chap. 3 for the Ioffe Drift area. Datum levels (first and last appearances) of zonal index species in individual cores were used to denote zonal boundaries. The oldest recovered sediments are of the Upper Pliocene, probably not older than ~4 Ma. The core correlation suggests a significantly reduced thickness of biostratigraphic zones in the sediment sections from the Ioffe Drift area compared to those in DSDP Site 516 from the neighboring Rio Grande Rise, indicating multiple erosional hiatuses. Moreover, up to five nannofossil and four foraminiferal zones are either washed out or mixed in some cores, supporting the occurrence of hiatuses that result in stratigraphic gaps.

1 Methods

The standard procedure was used to study the down-core species distribution and identify biostratigraphic zones and their boundaries (Ivanova et al. 2016, 2020). The dry sand fractions (>100 μm) from six cores (Table 1) were split to obtain aliquots containing >1000 planktic foraminiferal specimens and then viewed under the STEMI SV6 binocular microscope. These large aliquots were analyzed to better address the species levels of the first (FO) and/or last (LO) occurrence, which could be overlooked in commonly used aliquots of about 300 specimens. This special emphasis was used with species with well-established FO and/or LO datum levels. The sampling interval in the sediment cores varied from 5 to 30 cm, depending on sedimentological features, and it was usually 10 cm. Planktic foraminiferal specimens

E. Ivanova (✉) · O. Dmitrenko
Shirshov Institute of Oceanology, Russian Academy of Sciences, Moscow, Russia
e-mail: e_v_ivanova@ocean.ru

O. Dmitrenko
e-mail: senidol@yandex.ru

I. Murdmaa and E. Ivanova (eds.), *The Ioffe Drift*, Springer Geology,
https://doi.org/10.1007/978-3-030-82871-4_7

Table 1 Accelerator mass-spectrometry (AMS) radiocarbon dates from the Ioffe Drift and Rio Grande Rise. Dates from core AI-3318 after Ivanova et al. (2020). Poz–Poznan Radiocarbon Laboratory

Laboratory code	Core	Depth in core (cm)	Dated material	AMS-^{14}C date (years)
Poz-71478	AI-3318	10–11	Mixed planktic foraminifers	33,140 ± 580*
Poz-71479	AI-3318	20–21	Mixed planktic foraminifers	>46,000*
Poz-71475	AI-3320	5–6	Mixed planktic foraminifers	42,000 ± 2,000
Poz-71476	AI-3320	25–26	Mixed planktic foraminifers	27,350 ± 220
Poz-71467	AI-3321	10–11	Mixed planktic foraminifers	13,420 ± 70
Poz-71468	AI-3321	20–21	Mixed planktic foraminifers	22,050 ± 130

were analyzed from 46 samples in core AI-2436, 55 in core AI-3655, 50 in core AI-3316, 40 in core AI-3317, 34 in core AI-3318 and 30 in core AI-3319. Scanning electron microscope (SEM) photos of several stratigraphically important species were taken on TESCAN VEGA3 and VEGA-II XMU SEMs.

The qualitative abundance of calcareous nannoplankton taxa was established by the examination of smear-slides. Nannofossils were studied from four cores (all apart from AI-3655 and AI-3319) using an Amplival optical microscope (magnification 1,350) and SEM model JSM-U3 (magnification x6,000–10,000). Both microscopes were used for species identification and counting. SEM microphotos with a magnification of x2,000 were taken on TESCAN VEGA3 to identify tiny species. The down-core species distributions were studied in 44 raw samples in core AI-2436, 23 in core AI-3316, 47 in core AI-3317 and 24 in core AI-3318, generally from the same level where planktic foraminiferal specimens were identified.

2 Planktic Foraminiferal Assemblages and Stratigraphic Zones

2.1 *Assemblages*

Rich warm-water (subtropical to tropical) assemblages occur throughout all six records from the Ioffe Drift area. The upper parts of all cores contain the typical Quaternary species, with a diversity commonly ranging between 20 and 30 taxa per sample, apart from the intervals affected by dissolution. According to the taxonomy developed by Schiebel & Hemleben (2017), the extant species in the material studied include pink and white specimens of *Globigerinoides ruber* and *Globoturborotalita rubescens, Globigerinoides succucifer, Globorotalita tenella, Globigerinita glutinata, Neogloboquadrina dutertrei, Globorotalia menardii, Globorotalia inflata, Globigerinoides conglobatus, Globigerinella calida, Globigerinella siphonifera, Globoturborotalita tenella, Sphaeroidinella dehiscens, Neogloboquadrina incompta, Globorotalia crassaformis, Globorotalia truncatulinoides, Globorotalia hirsuta, Globigerina bulloides, Globigerina falconensis, Pulleniatina obliquiloculata, Orbulina universa,* and other species.

Several extinct Neogene–Quaternary species appeared down-core at various levels, including *Globorotalia crassaformis hessi, Globorotalia crassaformis viola, Globorotalia tosaensis, Globorotalia triangula, Globigerinoides fistulosus, Neogloboquadrina humerosa, Neogloboquadrina acostaensis, Globorotalia exilis, Globorotalia miocenica, Globorotalia multicamerata, Globorotalia margaritae, Pulleniatina primalis, Pulleniatina praecursor, Globigerina decoraperta, Globigerinoides obliquus extremus, Globorotalia pertenuis, Globorotalia limbata, Globorotalia puncticulata, Globorotalia plesiotumida, Dentoglobigerina altispira, Globoquadrina venezuelana, Sphaeroidinellopsis seminulina* and *Sphaeroidinellopsis kochi,* according to the taxonomy developed by Kennett and Srinivasan (1983) and Bolli and Saunders (1985).

SEM images of the most important taxa, including zonal index species, are provided in Plates 1, 2 and 3.

The same extant species were reported in the diverse assemblages of good-to-perfect preservation studied from both a shallow core, AI-3320, and a deeper core, AI-3321, collected respectively from the summit and slope of the Rio Grande Rise (see Table 1 in Chap. 4). In core AI-3320, rather well-preserved aragonite shells of some pteropod species were documented in Levels 0 and 35 cm, including *Styliola subula, Limacina inflata, Limacina bulimoides* and others.

It should be noted that upward microfossil reworking and/or down-core contamination is quite common in all cores studied. These phenomena are known to be typical also of the nearby Rio Grande Rise (Berggren et al. 1983a, b; Barash et al. 1983). Thus, we conclude that these phenomena have to be considered, at least, as regional, and the Ioffe Drift biochronostratigraphy must be developed with caution.

2.2 Foraminiferal Preservation

In all six cores studied from the Ioffe Drift, down-core foraminiferal and calcareous nannofossil preservation generally varies from perfect to moderate; however, several levels are affected by a significant selective dissolution resulting in poor preservation of calcareous microfossils (Figs. 1, 2, 3, 4, 5 and 6). Foraminiferal preservation is generally good to perfect on the drift summit, where weak dissolution was documented only in a few levels (145 cm in core AI-3655; 210–220 cm in core AI-3618; and 109, 207, 225, 350, 430–438, 600, 640–650 and 710 cm in core AI-2436). Dissolution is more common in the deeper cores from the drift slope. In core AI-3316,

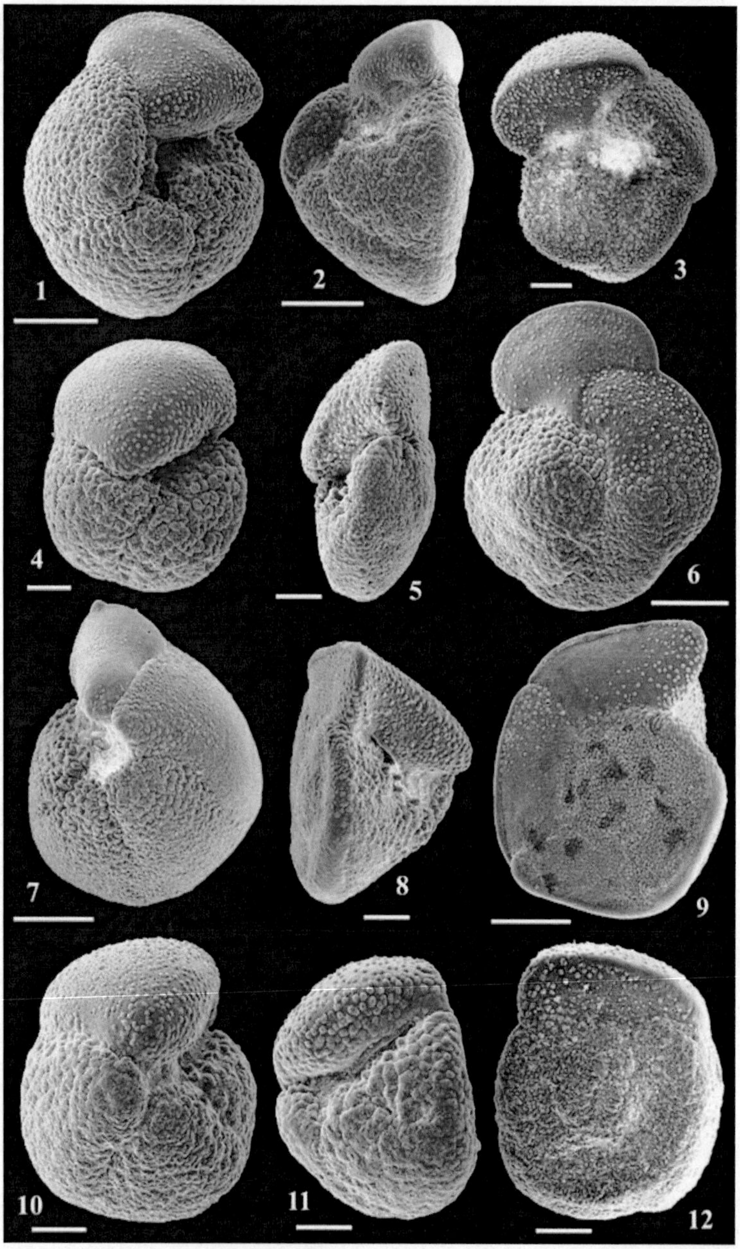

Plate 1 SEM images of planktic foraminiferal species: 1–3. Globorotalia crassaformis hessi, AI-3317, 330 cm, 1, 2 scale bar = 200 μm, 3 scale bar = 100 μm; 4–6. Globorotalia crassaformis viola, AI-3318, 290 cm, 4, 5 scale bar = 100 μm, 6 scale bar = 200 μm; 7–9. Globorotalia truncatulinoides, AI-2436, 286 cm, 7, 9 scale bar = 200 μm, 8 scale bar = 100 μm; 10–12. Globorotalia tosaensis, AI-3318, 210 cm, scale bar = 100 μm

Plate 2 SEM images of planktic foraminiferal species: 1–3. Globorotalia pertenuis, AI-3316, 300 cm, 1 scale bar = 500 μm, 2, 3 scale bar = 200 μm; 4–6. Globorotalia multicamerata, AI-2436, 660 cm, scale bar = 200 μm; 7–9. Globorotalia miocenica, AI-3317, 360 cm, scale bar = 100 μm; 10–11. Globigerinoides obliquus extremus, AI-2436, 660 cm, 10. scale bar = 200 μm, 11 scale bar = 100 μm; 12. Sphaeroidinellopsis kochi, AI-2436, 700 cm, scale bar = 200 μm

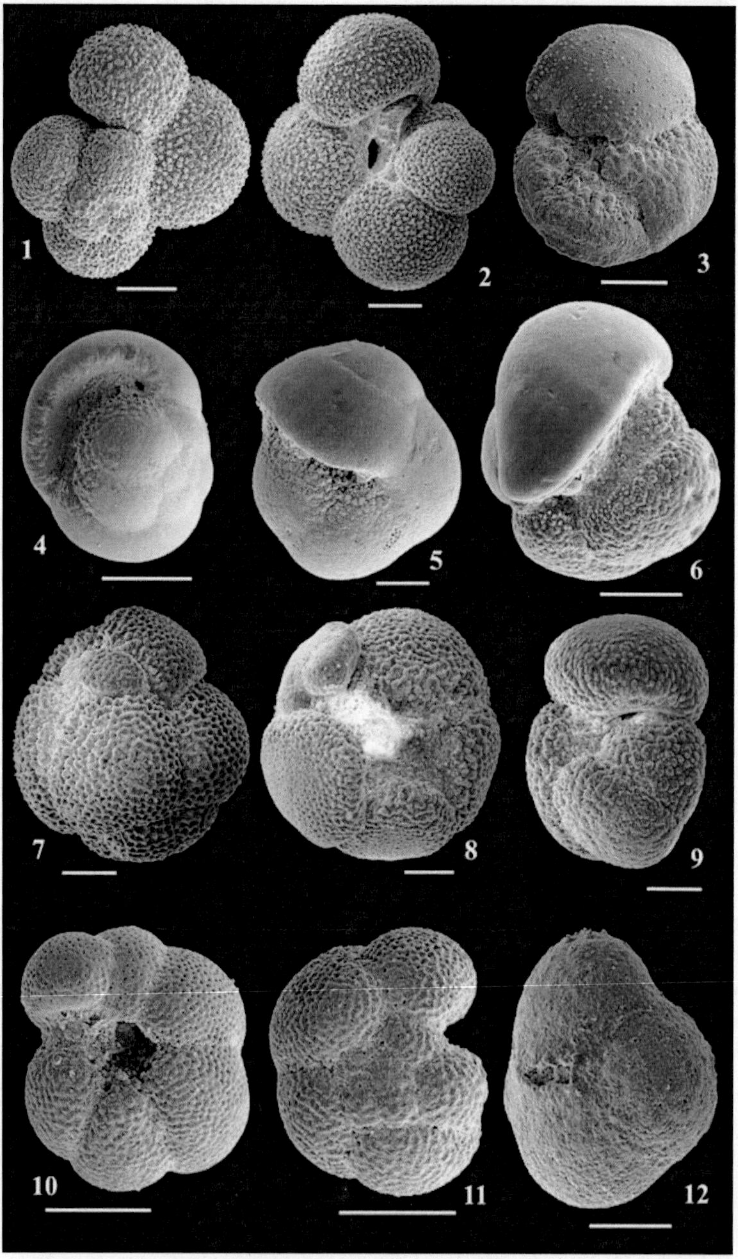

Plate 3 SEM images of planktic foraminiferal species: 1–3. Globigerinella calida calida, AI-3317, 125 cm, 1, 2 scale bar = 100 μm, 3. Globorotalia hirsuta, AI-2436, 400 cm, scale bar = 100 μm; 4. Pulleniatina praecursor, AI-3317, 512 cm, 4. scale bar = 200 μm; 5, 6. Pulleniatina primalis, 5. AI-3316, 420 cm, scale bar = 100 μm, AI-3317, 512 cm, scale bar = 200 μm; 7–9. Dentoglobigerina altispira, AI-3317, 512 cm, scale bar = 100 μm; 10, 11 Neogloboquadrina humerosa, AI-3316, 320 cm, scale bar = 200 μm; 12. Sphaeroidinellopsis seminulina, AI-3317, 512 cm, scale bar = 100 μm

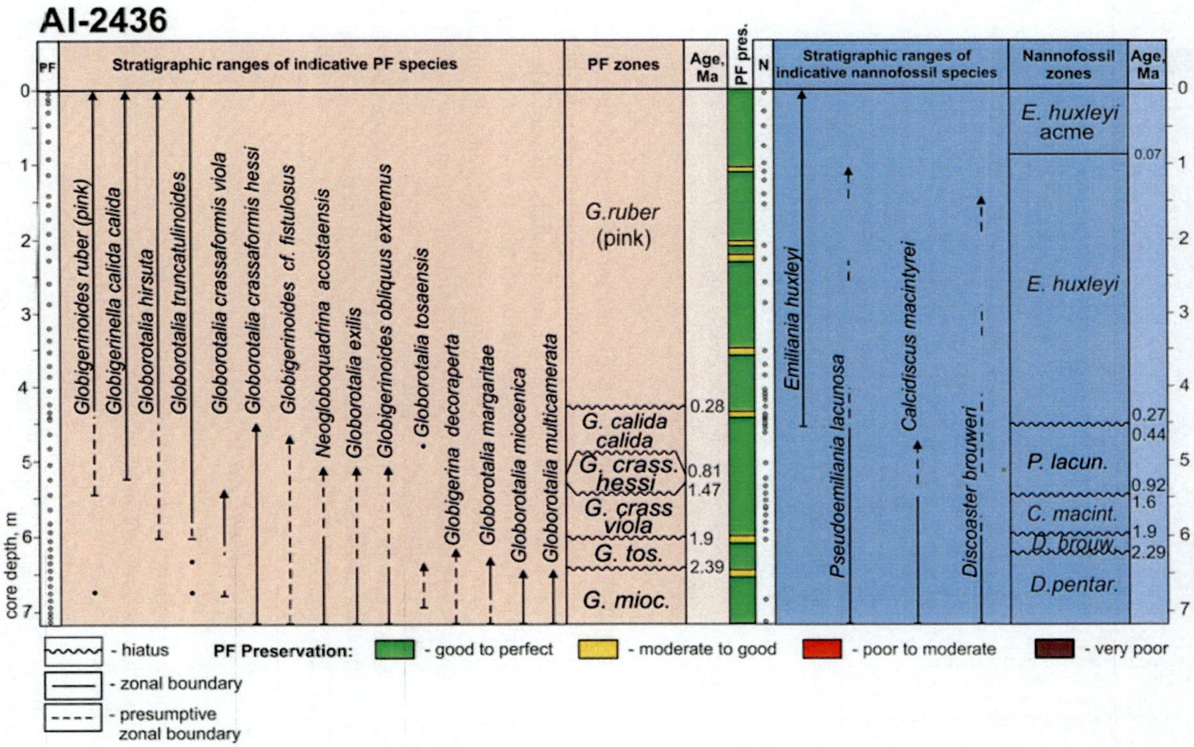

Fig. 1 Stratigraphy of core AI-2436: ranges of indicative species, planktic foraminiferal (PF) and nannofossil zones with their ages (modified from Ivanova et al. 2016). Samples studied for PF and nannofossils (N) are marked by dots in corresponding columns. Arrows indicate the last (presumably in situ) occurrence of species. See text for the species full names

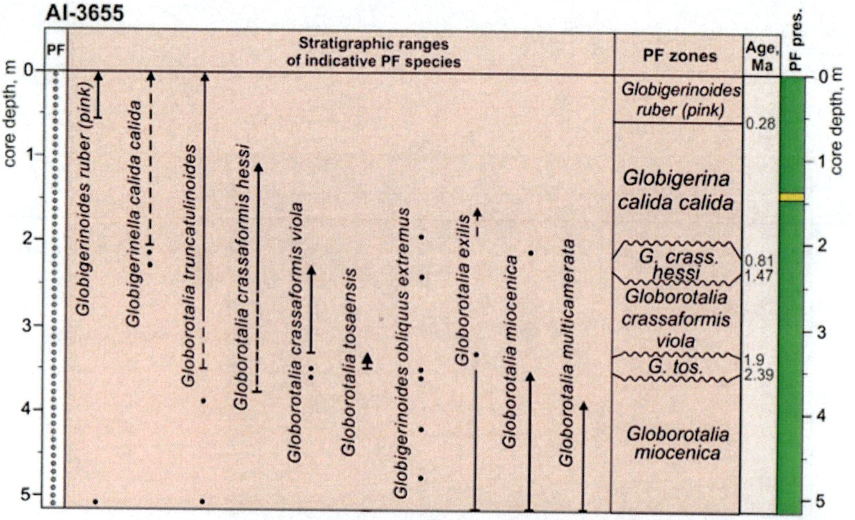

Fig. 2 Stratigraphy of core AI-3655: ranges of indicative species, planktic foraminiferal (PF) zones with their ages and foraminiferal preservation (PF pres.). Samples studied for PF are marked by dots in corresponding column. Arrows indicate the last (presumably in situ) occurrence of species. See text for the species full names and Fig. 1 for legend

good preservation dominates in the upper section and moderate is reported at 27, 47, 137, 147, 187, 257, 277, 307, 347 and 467 cm, while significant dissolution and poor preservation were noted at 377 and 407–427 cm. Similarly, in core AI-3317, significant dissolution was found in the lower part of the section, from 415 to 512 cm, while, upward, preservation was reported commonly as good, with some dissolution documented at a few levels including 25, 145, 155, 385 and 405 cm. In core AI-3319, collected to the WSW of the drift, moderate to heavy dissolution is characteristic of the lower part of the section, from 125 to 291 cm. Weaker dissolution was noted at 75–85 cm, while all other samples demonstrate rather good preservation.

2.3 *Planktic Foraminiferal Zones*

Despite the medium to poor down-core preservation in some sections, the occurrence of index species allowed the identification of planktic foraminiferal zones in all the cores investigated from the Ioffe Drift area (see Table 1 in Chap. 4). As a result, the foraminiferal zonation described in Chap. 3 (all units are referred to as "zones") was applied to stratify six sediment cores, as follows.

In core **AI-2436** from the drift summit (water depth 3,799 m, Fig. 1; Ivanova et al. 2016), which was collected the first and significantly stretched as a result of core processing (see Chap. 4), the uppermost zone was that of *Globigerinoides ruber* (pink) (430–0 cm, 0.28–0 Ma), with the persistent occurrence of the index species.

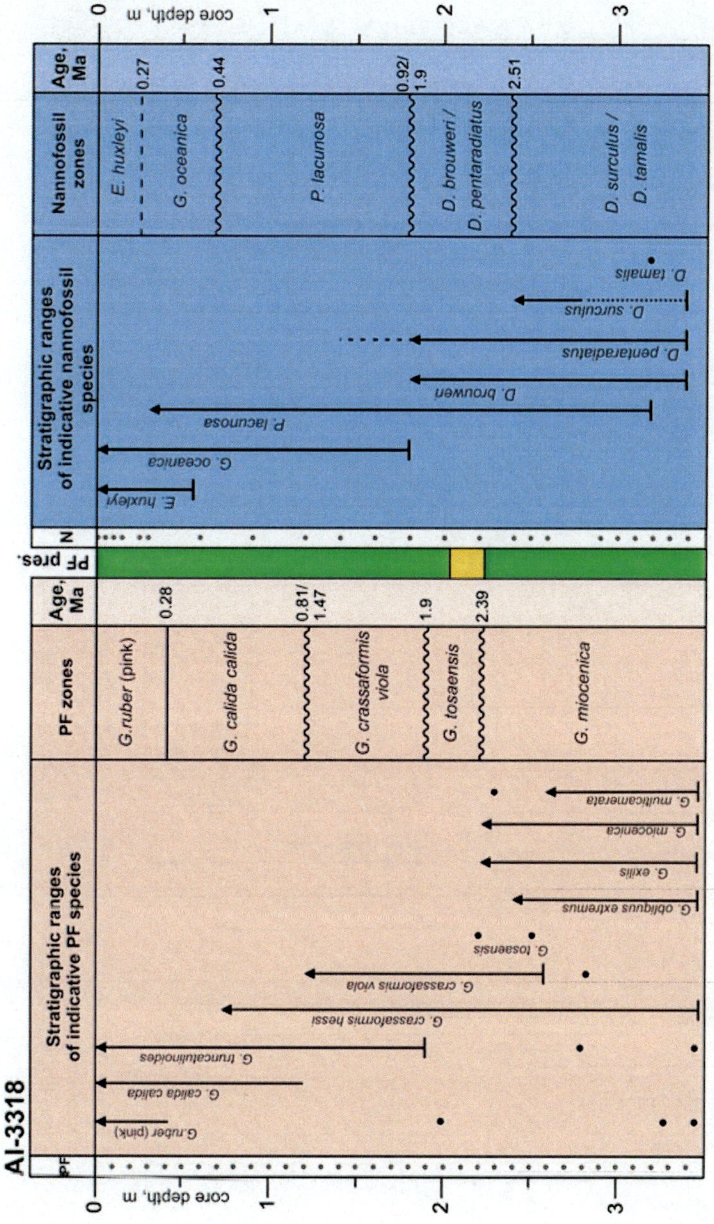

Fig. 3 Stratigraphy of core AI-3318: ranges of indicative species, planktic foraminiferal (PF) and nannofossil zones with their ages (modified from Ivanova et al. 2020) and foraminiferal preservation (PF pres.). Samples studied for PF and nannofossils (N) are marked by dots in corresponding columns. Arrows indicate the last (presumably in situ) occurrence of species. See text for the species full names and Fig. 1 for legend

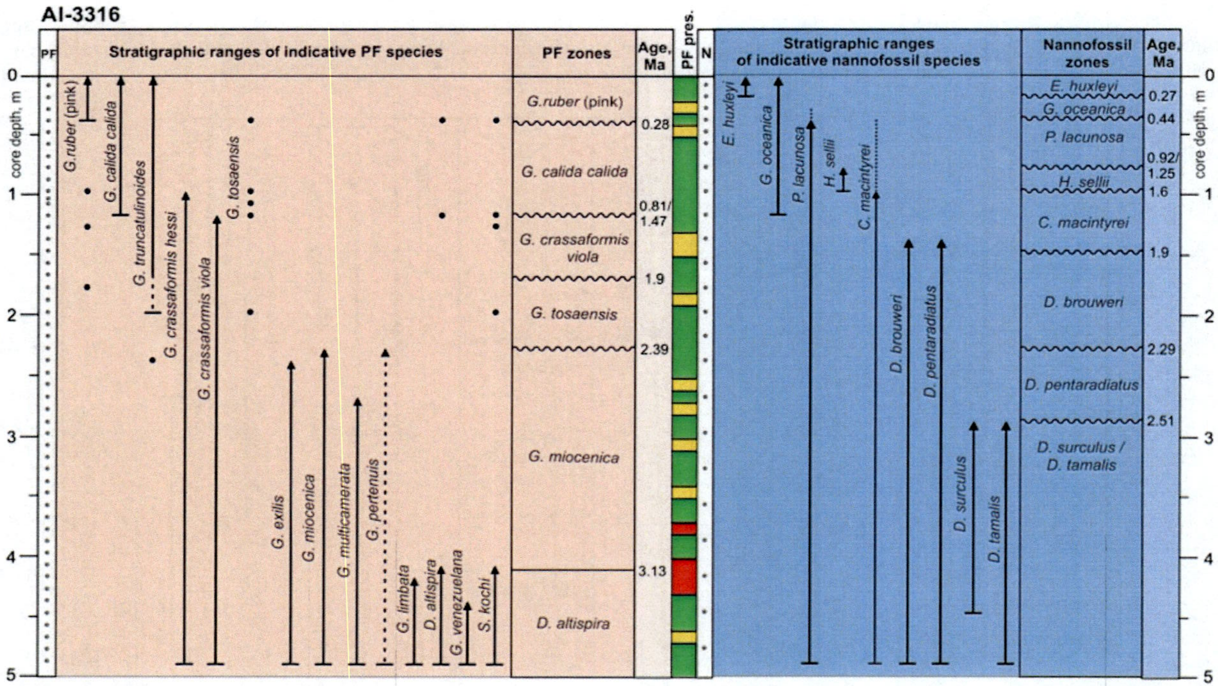

Fig. 4 Stratigraphy of core AI-3316: ranges of indicative species, planktic foraminiferal (PF) and nannofossil zones with their ages (modified from Ivanova et al. 2020) and foraminiferal preservation (PF pres.). Samples studied for PF and nannofossils (N) are marked by dots in corresponding columns. Arrows indicate the last (presumably in situ) occurrence of species. See text for the species full names and Fig. 1 for legend

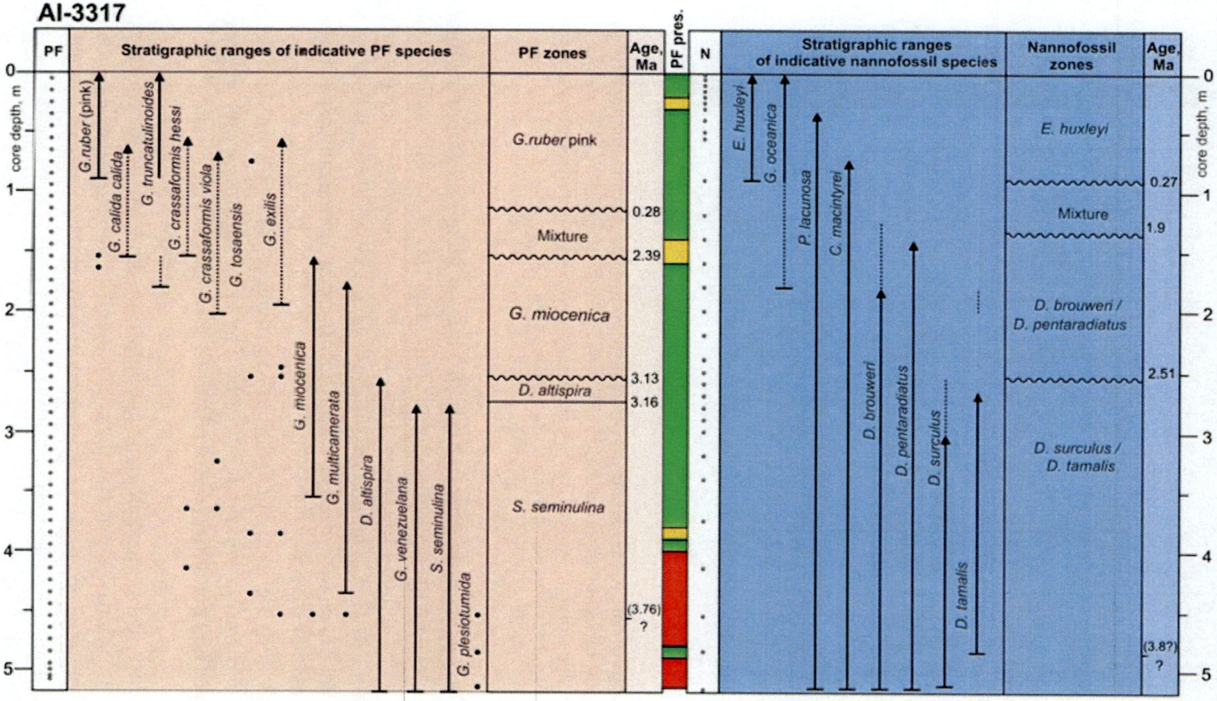

Fig. 5 Stratigraphy of core AI-3317: ranges of indicative species, planktic foraminiferal (PF) and nannofossil zones with their ages and foraminiferal preservation (PF pres.). Samples studied for PF and nannofossils (N) are marked by dots in corresponding columns. Arrows indicate the last (presumably in situ) occurrence of species. See text for the species full names and Fig. 1 for legend

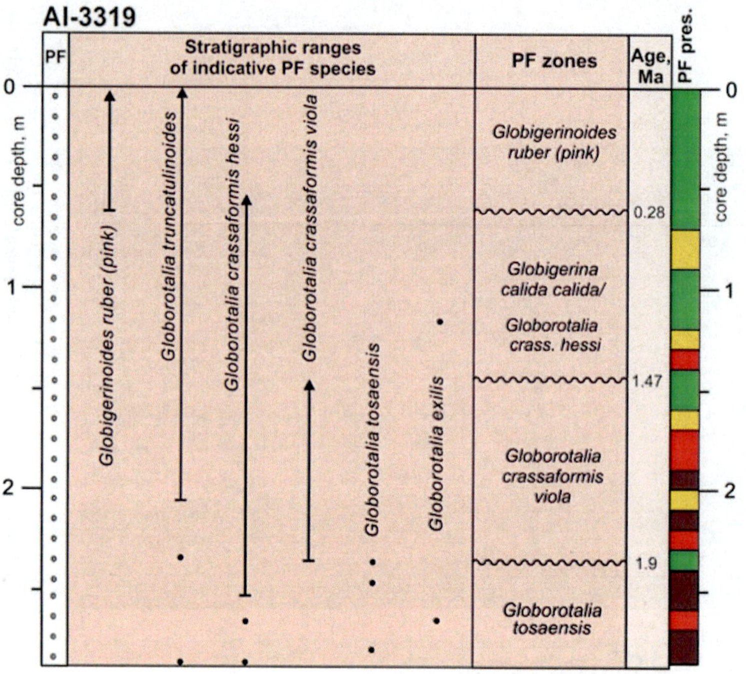

Fig. 6 Stratigraphy of core AI-3319: ranges of indicative species, planktic foraminiferal (PF) zones with their ages and foraminiferal preservation (PF pres.). Samples studied for PF are marked by dots in corresponding column. Arrows indicate the last (presumably in situ) occurrence of species. See text for the species full names and Fig. 1 for legend

Globigerinella calida calida zone (0.81–0.28 Ma) was identified within the interval 520–430 cm, based on the occurrence of the nominate taxon as well as *Globorotalia crassaformis hessi* and *G. truncatulinoides.* The lower boundary of the underlying rudimentary zone of *Globorotalia crassaformis hessi* (530–520 cm, 1.47–0.81 Ma) is marked by the LO of *Globorotalia crassaformis viola,* an indicative species of the underlying same-named zone (600–530 cm, 1.9–1.47 Ma). The upper boundary of an also reduced underlying zone of *Globorotalia tosaensis* (640–600 cm) was defined by the FO of *G. truncatulinoides* at ~1.9 Ma. The latter species occurs persistently within the upper 600 cm of the core, whereas *G. tosaensis* occurs randomly at a few levels.

The boundary between zones of *Globorotalia tosaensis* and *Globorotalia miocenica* were inferred from the disappearance of Neogene species including *Globorotalia multicamerata* and *G. miocenica* at 640 cm. As mentioned in Chap. 3, the latter datum level was adopted from Wade et al. (2011), namely the LO of *G. miocenica* at about 2.39 Ma. Although the LO of *G. multicamerata* is suggested to be older, ~3 Ma (Wade et al. 2011, see Chap. 3), in core AI-2436 both species disappeared simultaneously, probably due to the strongly reduced thickness of the stratigraphic zones. The lowermost part of the core (714–640 cm) is referred to as the recovered part of *Globorotalia miocenica* zone on the basis of the occurrence of nominate taxon, *Globigerina decoraperta, Sphaeroidinellopsis seminulina, Globorotalia exilis, Globigerinoides obliquus extremus, Neogloboquadrina acostaensis, Globorotalia multicamerata* and *Globigerinoides fistulosus.* The assumption about the Upper Pliocene character of foraminiferal fauna is corroborated by the absence of older Pliocene species, such as *Globoquadrina dehiscens* and *Dentoglobigerina altispira,* and of the above-mentioned indicative Pleistocene species. The reduced thickness of foraminiferal zones in the core section indicates the occurrence of several hiatuses, especially pronounced in *Globorotalia crassaformis hessi* zone.

Another core, AI-3655, was collected from the drift summit (water depth 3,799 m) in order to overcome the problems for the stratigraphy caused by core AI-2436 stretching. Core **AI-3655** recovered the same foraminiferal zones, according to the down-core distributions of the nominate taxa (Fig. 2). The *Globigerinoides ruber* (pink) zone is only 58 cm thick (as it was not stretched), whereas the underlying zone of *Globigerinella calida calida* (218–58 cm) is considerably thicker than in core AI-2436; however, its lower boundary is doubtful, like that of the underlying *Globorotalia crassaformis hessi* zone (230–218 cm). These two are challenging to identify in both cores due to the rare occurrence of a species susceptible to dissolution: *G. calida calida.* Therefore, the boundary of the *Globorotalia crassaformis hessi* and *Globorotalia crassaformis viola* zones (330–230 cm), and the thickness of the latter, are questionable. The *Globorotalia tosaensis* zone (350–330 cm) is strongly reduced, whereas the lowermost recovered zone of *Globorotalia miocenica* (514–350 cm) is rather well represented in core AI–3655. Several hiatuses in the core section are suggested by the reduced thickness of most foraminiferal zones.

In core **AI-3318** from the drift summit (water depth 3,788 m, Ivanova et al., 2020), pink specimens of *Globigerinoides ruber* occur continuously in the upper 40 cm, defining the self-titled uppermost foraminiferal zone (Fig. 3). The absence of Holocene sediments and the Late Pleistocene age of the upper 20 cm of the core section were ascertained by two AMS-^{14}C dates of 33,140 $\pm$ 580 ka and >46,000 ka BP, obtained at 10 and 20 cm, respectively (Table 1). Scarce tests of *G. ruber* were found at 192, 326 cm and in the core-catcher (CC), probably as a result of core processing and/or down-core contamination. Based on the down-core distribution of index species, the zone of *Globigerinella calida calida* (0.81–0.28 Ma) is tentatively identified as between 120 and 40 cm, while *Globorotalia crassaformis hessi* zone (1.47–0. 81 Ma) is probably washed out, suggesting a hiatus at 120 cm, where both the FO of *G. calida calida* and the LO of *G. viola* were reported at the same level. The base of the underlying *Globorotalia crassaformis viola* zone (1.9–1.47 Ma) is defined by the persistent occurrence of *G. truncatulinoides* above 190 cm. The *Globorotalia tosaensis* zone (2.39–1.9 Ma) bounded below by the LO of *Globorotalia miocenica* covering the interval 220–190 cm, possibly with both lower and upper boundaries represented by hiatuses in line with the comparison with magnetic susceptibility data (see Chapter 9). Some extinct Pliocene species notably *Globorotalia exilis, G. multicamerata* and *Globigerinoides obliquus extremus,* disappear in *Globorotalia miocenica* zone (340–220 cm) or PL5 (after Berggren et al. 1995). Single specimens of *Globorotalia pertenuis, G. margaritae* and other extinct species were also documented in this zone. The *Globorotalia miocenica* zone seems not to be completely recovered, as the LO of *Dentoglobigerina altispira* (3.13 Ma, according to Wade et al. 2011) is not identified.

The same and some older extinct taxa are found in core **AI-3316** from the northern slope of the drift, which penetrated stratigraphically deeper than the three cores from the drift summit (water depth 3,898 m, Ivanova et al. 2020). In the core section, foraminiferal zones are defined according to the nominate taxa distribution, as follows: *Globigerinoides ruber* (pink) from 37 to 0 cm, *Globigerinella calida calida* from 117 to 37 cm, *Globorotalia crassaformis viola* from 167–117 cm and *Globorotalia tosaensis* from 227 to 167 cm (Fig. 4). The *Globorotalia crassaformis hessi* zone is not preserved in the core section and a hiatus is assumed at 117 cm. The *Globorotalia tosaensis* zone seems also to be bounded by hiatuses (see Chap. 9). The *Globorotalia miocenica* (or PL5) zone is documented between the LO of the same named species and the LO of *Dentoglobigerina altispira,* at 227 cm and 407 cm, respectively. The lowermost *Dentoglobigerina altispira* zone (or PL4, 3.16–3.13 Ma, 489–407 cm) contains common specimens of Neogene species along with index species, notably *Globorotalia limbata, Globoquadrina venezuelana* and *Sphaeroidinellopsis kochi.* According to Kennett and Srinivasan (1983), the latter two species are typical of the tropical early Pliocene zone N19, after Blow (1969). They were probably reworked upward. A significant dissolution of foraminiferal shells was reported from two intervals of the lowermost zones of *Globorotalia miocenica* and *Dentoglobigerina altispira.*

The most stratigraphically biased core **AI-3317** from the northern slope (water depth 3,832 m) probably penetrated deeper into the Pliocene section than any other core. It is characterized by rather significant contamination and reworking, which are probably stronger than elsewhere. The uppermost *Globigerinoides ruber* (pink) zone is identified from 115 to 0 cm by the persistent occurrence of the zonal marker (Fig. 5); however, the underlying 40 cm interval contains a mixture of several index species and probably overlies a hiatus. The sediments of the *Globigerinella calida calida, Globorotalia crassaformis hessi, Globorotalia crassaformis viola* and *Globorotalia tosaensis* zones are washed out by bottom-water erosion, with only mixed remnants in the 155–115 cm interval. The concurrent presence of the abovementioned taxa suggests some hiatuses within this interval. By contrast, the underlying zones of *Globorotalia miocenica* (255–155 cm) and *Dentoglobigerina altispira* (275–255 cm), respectively PL5 and PL4 (after Berggren et al. 1995), are well documented. Nevertheless, the occurrence of hiatuses cannot be denied within these zones and at their boundaries, especially considering the magnetic susceptibility data (see Chaps. 6 and 9). The *Sphaeroidinellopsis seminulina* zone or PL3 (3.84–3.16 Ma, after Wade et al. 2011) was reported below, within the interval from 255 cm down to at least 455 cm, according to the occurrence and the LO of the nominate taxon. The LO of *Globorotalia plesiotumida* (3.76 Ma, according to Wade et al. 2011) is suggested at 455 cm. The lower part of the zone is characterized by poor foraminiferal preservation due to significant dissolution at several levels.

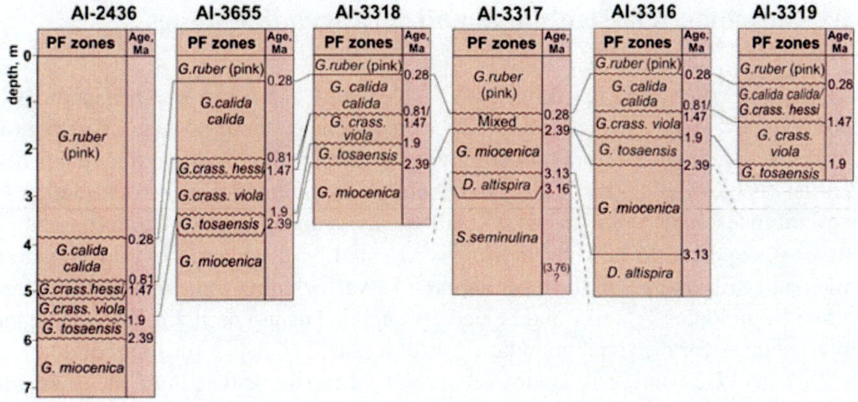

Fig. 7 Correlation of planktic foraminiferal zones in six cores studied. See text for the species full names and Fig. 1 for legend

The deepest and shortest core, **AI-3319,** collected to the WSW of the Ioffe Drift from a water depth of 4,066 m, did not recover the Pliocene/Pleistocene boundary (Fig. 6). Foraminiferal fauna from the lower part of the core, below 170 cm, were heavily affected by dissolution therefore some samples contain only a few whole foraminiferal tests. Nonetheless, the resistance to dissolution of some species, notably globorotaliids, allows the definition of the zones and their boundaries. The *Globigerinoides ruber* (pink) zone, with persistent occurrence of the nominate taxon, was identified from 60 to 0 cm. It is underlined by a combined zone of *Globigerinella calida calida/Globorotalia crassaformis hessi*, probably below a hiatus at ~60 cm. A better-represented *Globorotalia crassaformis viola* zone seems to be bounded by hiatuses at 140 and 235 cm. The lowermost *Globorotalia tosaensis* zone was recovered at an interval of 288–235 cm, demonstrating a visible thickness of 53 cm.

The stratigraphic correlation of foraminiferal zones in the six cores studied from the Ioffe Drift is shown in Fig. 7. This figure documents the best representation of Pleistocene zones in cores AI-2436 and AI-3655, from the drift summit. By contrast, cores AI-3316 and AI-3317 from the northern slope recovered the older sediments of the Upper Pliocene (>3.13 Ma and ~4 Ma, respectively). Correspondingly, the stratigraphically oldest sediments of the *Sphaeroidinellopsis seminulina* zone were recovered only by core AI-3317. In four cores, the *Globigerinella calida calida* and *Globorotalia crassaformis hessi* zones are missing or cannot be properly determined, and are therefore combined. At the same time, core AI-3317 contains a condensed interval of mixed faunal zones, demonstrating the strongest contamination, reworking and erosion.

3 Nannofossil Assemblages and Stratigraphic Zones

In the four cores from the Ioffe Drift (Table 1 in Chap. 4) in which nannofossils were studied, they were common and represented by about 40 taxa. Nannofossil zones were recognized from four cores and the data for three of them were published earlier (Ivanova et al. 2016, 2020). It should be mentioned that the distribution of several index species, for instance small *Gephyrocapsa* (mostly *G. sinuosa* and *G. aperta*), appeared to have been strongly affected by dissolution in corresponding intervals. SEM images of the most important taxa, including zonal index species, are provided in Plates 4, 5, 6, 7 and 8. Here, we applied nannofossil zonation, using the revised dates for the zonal boundaries, as described in Chap. 3 for four cores.

In core **AI-2436** (Fig. 1; Ivanova et al. 2016), from the drift summit, the following Pliocene–Quaternary nannofossil zones were inferred downcore: *Emiliania huxleyi* zone, 452–0 cm (0.27–0 Ma), including the *E. huxleyi* Acme zone at 87–0 cm (0.07–0 Ma); *Gephyrocapsa oceanica* zone (0.44–0.27 Ma) is probably washed out and has been partly reworked upward; while the interval 548–452 cm delineates *Pseudoemiliania lacunosa* zone (0.92–0.44 Ma). The small *Gephyrocapsa* zone (1.25–0.92 Ma) was not found, while the *Helicosphaera sellii* zone (1.6–1.25 Ma) was probably washed out and its index species reworked into the *E. huxleyi* zone. The zone of *Calcidiscus macintyrei* (1.9–1.6 Ma) is defined by the interval 595–548 cm, above the persistent occurrence of *Discoasters*. The lowermost part of the core (714–595 cm) consists of two zones: a rudimentary *Discoaster brouweri* zone (620–595 cm, 2.29–1.9 Ma); and a recovered part of zone *Discoaster pentaradiatus* (714–620 cm). The absence of some zones and the rudimentary character of the others clearly point to hiatuses, at least at 452 cm (0.44/0.27 Ma), 548 cm (1.51/0.92 Ma), 595 cm (1.9 Ma) and 620 cm (2.29 Ma).

In core **AI-3318** (Figs. 3 and 8; Ivanova et al. 2020) from the drift summit, *Emiliania huxleyi* occurs in the upper 55 cm and its upward-increasing abundance defines the lower boundary (0.27 Ma) of the uppermost self-titled zone, presumably at 25 cm. Other index species, *Gephyrocapsa oceanica* and *Pseudoemiliania lacunosa,* were reported from the intervals 180–0 cm and 340–25 cm, respectively. The boundary between the zones of these names at 0.44 Ma is tentatively located at 70 cm on the basis of the increasing abundance of *G. oceanica* and decreasing abundance of *P. lacunosa,* and the hiatus is very likely to have occurred at this level. The lower boundary of the *Pseudoemiliania lacunosa* zone (0.92 Ma), represented by a hiatus, is tentatively placed at 180 cm, whereas the three underlying zones of Gartner's scheme (1977), notably the zones of small *Gephyrocapsa* (1.25–0.92 Ma), *Helicosphaera sellii* (1.6–1.25 Ma) and *Calcidiscus macintyrei* (1.9–1.6 Ma), seem to have been washed out. Various *Discoasters* occur persistently down-core below 180 cm, while specimens of *D. pentaradiatus* are probably contaminated upwards to 140 cm. The existing limits to resolution allow no greater refinement than the combined zones of *Discoaster brouweri/Discoaster pentaradiatus* (2.51–1.9 Ma) and of *Discoaster surculus/Discoaster tamalis* (2.51–2.73 Ma) within the intervals of 240–180 cm and 345–240 cm, respectively. The LO of *D. surculus* defines the boundary between

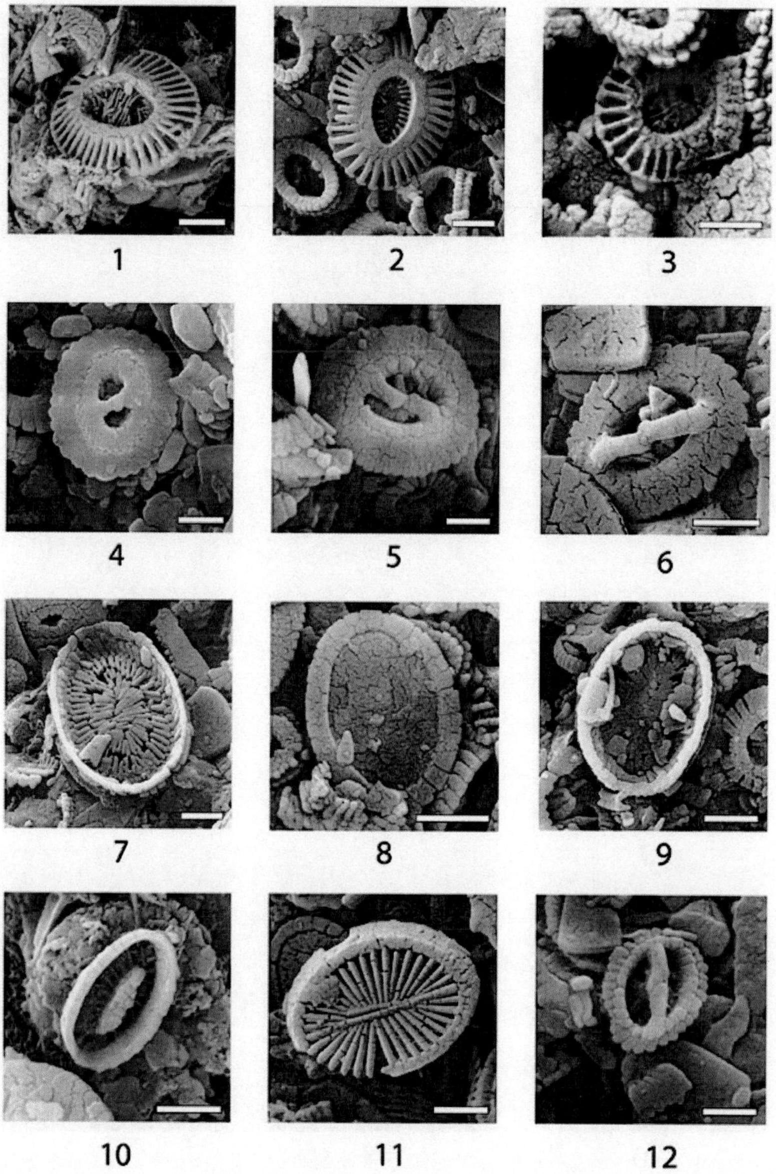

Plate 4 SEM images of nannofossils from Core AI–3318: 1–3. Emiliania huxleyi (Lohm.) Hay, Mohl., scale bar = 1 μm; 1–55, 2–20, 3–180 cm. 4. Gephyrocapsa oceanica Kpt., 280 cm, scale bar = 1 μm. 5,6. Gephyrocapsa caribbeanica Boudr., Hay, scale bar = 1 μm; 5–340, 6–0 cm. 7, 9. Syracosphaera pulchra Lohm., 7–15 cm, scale bar = 1 μm, 9–300 cm, scale bar = 2 μm. 8. Syracosphaera cf. pulchra Lohm., 290 cm, scale bar = 2 μm. 10. Syracosphaera ossa (L.-Schl.) Loedl., Tapp., scale bar = 1 μm. 11. Syracosphaera sp., scale bar = 1 μm. 12. Gephyrocapsa aperta Kpt., 50 cm, scale bar = 1 μm

Plate 5 SEM images of nannofossils from Core AI–3318: Core AI–3318. 1. Gephyrocapsa protohuxleyi McIntyrei, 50 cm, scale bar = 1 μm. 2–4. Acanthoica quattrospina Lohm., scale bar = 2 μm. 2, 3–280 cm. 4–300 cm. 5. Umbilicosphaera sibogae (Weber van Bosse) Gaarder., 10 cm. 6. U. anulus (Lecal) Young, Geisen, 15 cm, scale bar = 2 μm. 7. Calcidiscus leptoporus (Murr., Blackm.) Loebl., 0 cm, scale bar = 5 μm. 8. Discosphaera tubifera (Murr., Blackm.) Kpt., 0 cm. 9, 10. Umbellosphaera irregularis Paasche; 55 cm, scale bar = 2 μm. 11, 12. Umbellosphaera tenuis (Kpt.) Paasche, scale bar = 2 μm; 11–5, 12–0 cm

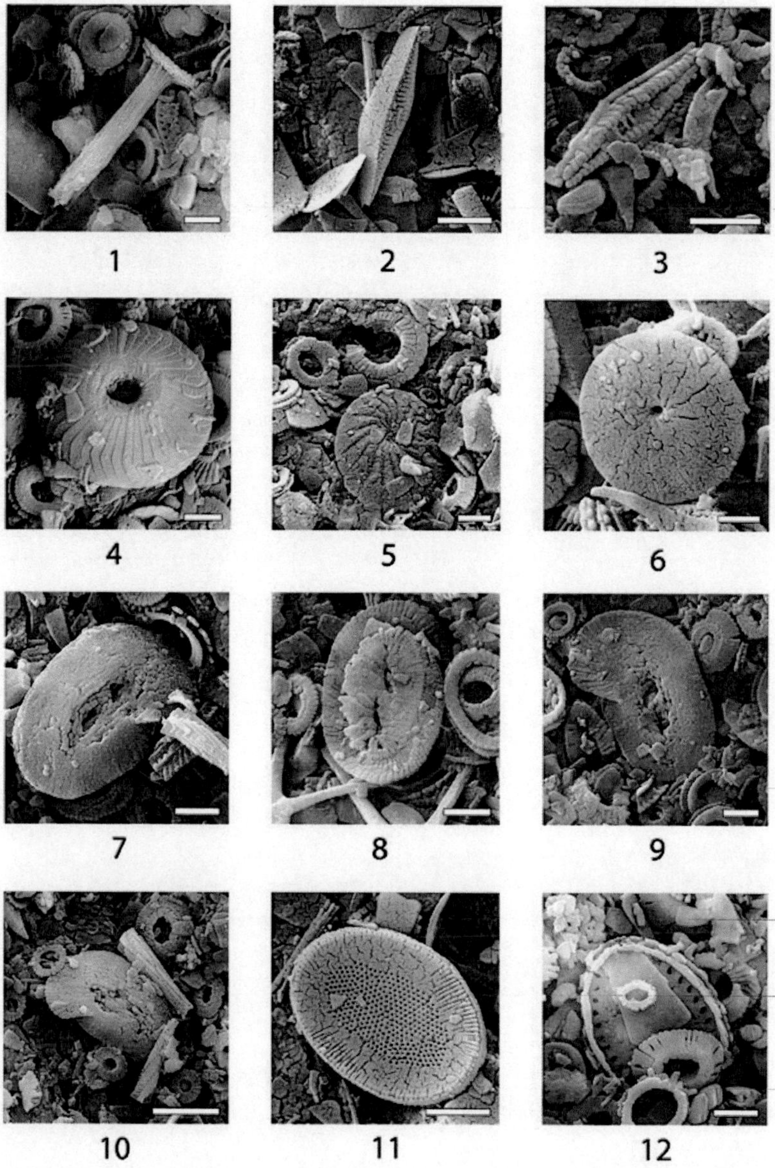

Plate 6 SEM images of nannofossils from the Ioffe Drift sections: 1. Rhabdosphaera clavigera (Murr., Blackm.) Kpt., 140 cm, scale bar = 2 μm. 2, 3. Scapholithus fossilis Defl., scale bar = 2 μm, 2–0, 3–120 cm. 4. Calcidiscus macintyrei (Bukry, Bram.) Loebl., Tapp., 160 cm, scale bar = 2 μm. 5, 6. Oolithothus antillarum (Cohen) Rein, 5–25 cm, scale bar = 2 μm, 6–160 cm, scale bar = 1 μm. 7. Helicosphaera carteri (Wall.) Kpt., 90 cm, scale bar = 2 μm. 8. Helicosphaera wallichii (Lohm.) Boudr., Hay, 55 cm, scale bar = 2 μm. 9. Helicosphaera sellii Bukry, Bram., 260 cm, scale bar = 2 μm. 10. Helicosphaera hyalina Gaarder, 60 cm, scale bar = 5 μm. 11. Pontosphaera japonica (Tak.) Nish. 0 cm, scale bar = 2 μm. 12. Pontosphaera multipora (Kpt.) Roth., 140 cm, scale bar = 2 μm

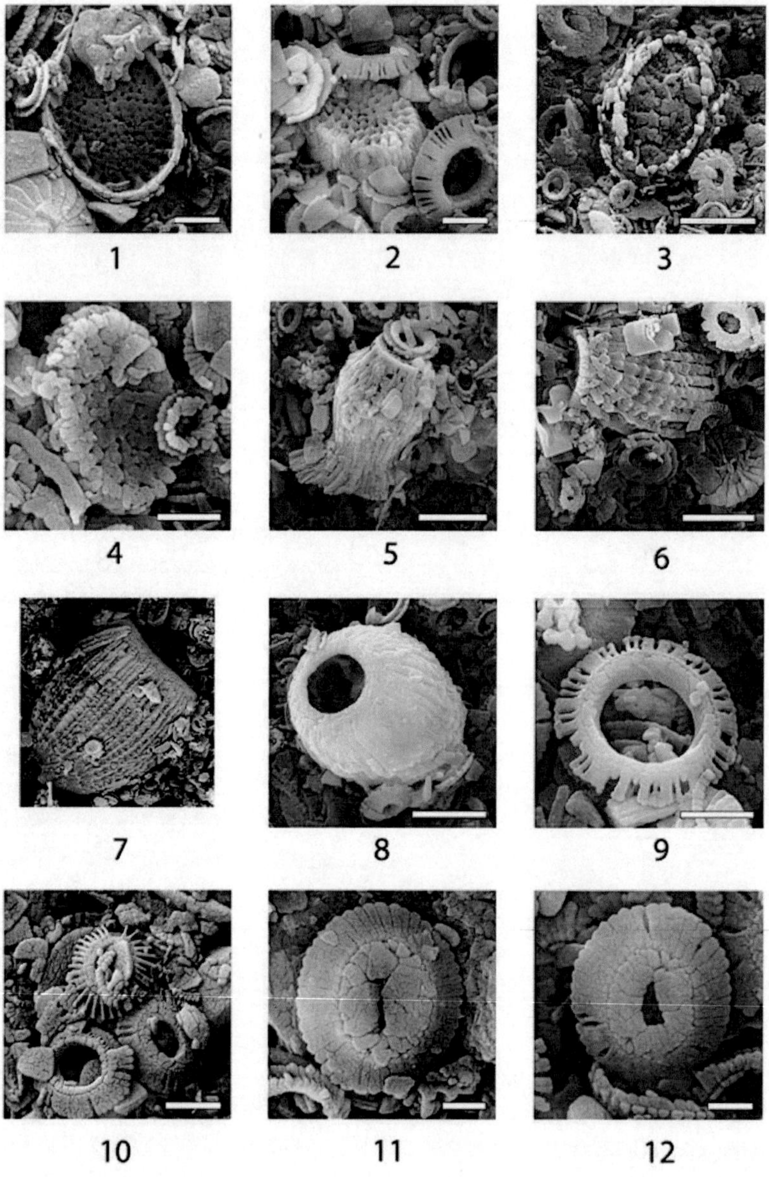

Plate 7 SEM images of nannofossils from Core AI–3318: 1, 2. Pontosphaera multipora (Kpt.) Roth., scale bar = 2 μm, 1–140 cm, 2– 45 cm. 3, 4. Pontosphaera segmenta (Bukry, Perciv.), 3–45 cm, scale bar = 5 μm, 4–340 cm, scale bar = 2 μm. 5. Scyphosphaera amphora Defl., 220 cm, scale bar = 5 μm. 6, 7. Scyphosphaera aranta Kpt.; 6–310 cm, scale bar = 5 μm, 7–60 cm. 8. Scyphosphaera globulata Bukry, Perciv, 45 cm, scale bar = 5 μm. 9, 10. Pseudoemiliania lacunosa (Kpt.) Gart.; scale bar = 2 μm, 9–140, 10–180 cm. 11, 12. Reticulofenestra pseudoumbilica (Gart) Gart., 11, 12–210 cm, scale bar = 1 μm

Plate 8 SEM images of nannofossils from the Ioffe Drift sections: Core AI–3318:1. Ceratolithus cristatus Kpt., 55 cm, scale bar = 5 μm. 2,3. Ceratolithus tricorniculatus Gart., 2–90 cm, scale bar = 5 μm, 3–250 cm, scale bar = 10 μm. 4. Ceratolithus acutus Gart., Bukry, 250 cm, scale bar = 5 μm. 5. Ceratolithus rugosus Bukry, Bram., 300 cm, scale bar = 5 μm. 6. Discoaster tamalis Kpt., 310 cm, scale bar = 2 μm. 7–9. Discoaster surculus Mart., Bram; 7–300 cm, scale bar = 2 μm, 8–280 cm, scale bar = 2 μm, 9–290 cm, scale bar = 10 μm, 9 (left) Discoaster calcaris Gart., 10. Discoaster moorei Bukry and Discoaster calcaris Gart., 250 cm, scale bar = 10 μm. Core AI–3316: 11 (left) Discoaster brouweri Tan., (right) Discoaster tamalis Kpt., 197 cm, scale bar = 5 μm. 12. Discoaster pentaradiatus Tan., 267 cm, scale bar = 5 μm

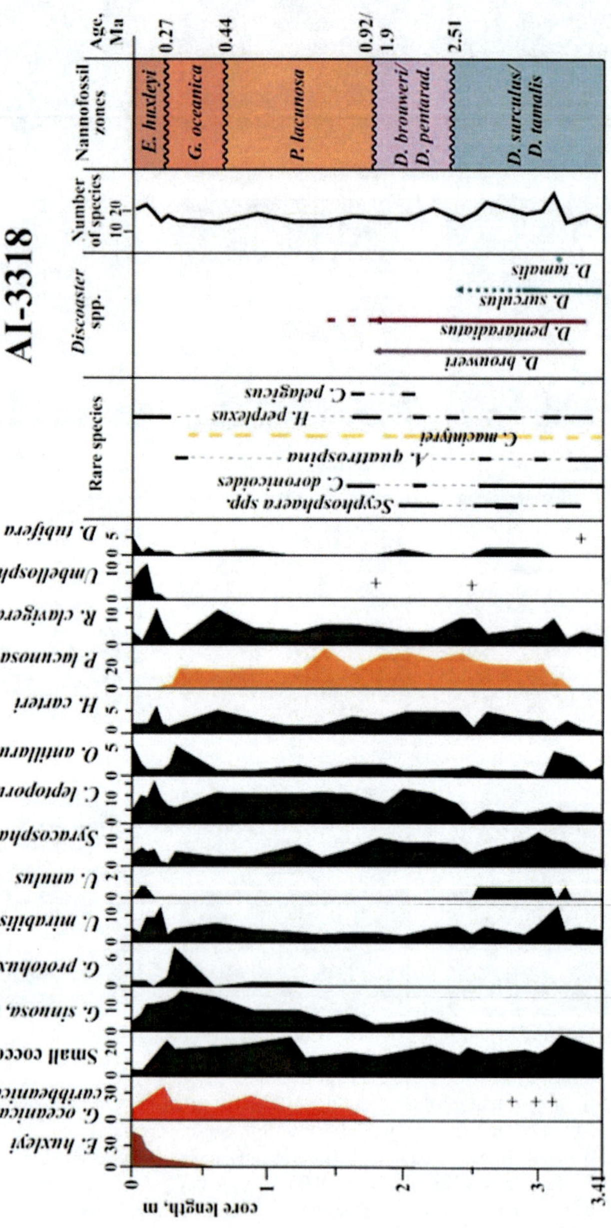

Fig. 8 Down-core distribution of species and nannofossil zones from core AI-3318. See text for the species full names and Fig. 1 for legend

these zones at 240 cm. *D. tamalis* was found in only one sample, from 320 cm. All zonal boundaries in the core, perhaps excepting the uppermost (0.27 Ma), seem to be represented by hiatuses (see also Chap. 9).

In core **AI-3316** (Figs. 4 and 9; Ivanova et al. 2020), from the northern slope of the drift, the presence of *Emiliania huxleyi* in one sample from 7 cm gives the lower erosional boundary of the reduced zone of this same name. Similarly, the other Quaternary zones are strongly reduced, as follows, from the down-core distribution of the zonal markers *Gephyrocapsa oceanica* and *Pseudoemiliania lacunosa* within the intervals 177–7 cm and 489–27 cm, respectively. The boundary between the same-named zones (0.44 Ma) is defined by the last persistent occurrence of the latter species at 37 cm and definitely represents a hiatus. The latter is also true for the boundary between the *Pseudoemiliania lacunosa* (0.92–0.44 Ma) and *Helicosphaera sellii* (1.6–1.25 Ma) zones, at 77 cm, as the small *Gephyrocapsa* zone (1.25–0.92 Ma) is washed out. An interval 97–77 cm, marked by the presence of *H. sellii,* delineates the zone of this name. Although the nominate taxon of the underlying zone *Calcidiscus macintyrei* (1.9–1.6 Ma) occurs up to 37 cm, large coccoliths of the taxa were documented only within the interval 488–97 cm, thereby assigning the boundary of zones *Helicosphaera sellii* and *Calcidiscus macintyrei* to 97 cm. The disappearance of *Discoasters* determines the boundary of the latter zone and the *Discoaster brouweri* zone at 137 cm. Both *D. brouweri* and *D. pentaradiatus* occur downcore below this level; however, the boundary between the same-named zones (2.29 Ma, according to Bergen et al. 2019) could be identified by an upward increase in the latter species abundance up to 227 cm. By contrast, only the combined *Discoaster surculus/Discoaster tamalis* zone (>2.73–2.51 Ma) could be defined in the lower part of the section, from 489 to 287 cm, based on the species distribution within the intervals 448–297 cm and 489–297 cm, respectively. The rather persistent occurrence of *D. tamalis* in the latter interval, together with the absence of *D. surculus* from its lowermost part, suggests that the core recovered at least 2.73 Ma, according to available publications on the datum levels of these species (i.e., Rahman & Rothe, 1989; Backman et al., 2012; Bergen et al., 2019). All zonal boundaries in the core are probably represented by hiatuses (see also Chap. 9).

In the most biased core, **AI-3317**, from the northern slope of the Ioffe Drift (Figs. 5 and 10), the *Emiliania huxleyi* zone is defined as the interval with the persistent occurrence of nominate taxon, at 90–0 cm, while the underlying interval (137–90 cm) contains a mixture of several species and cannot be attributed to any specific nannofossil zone. It is separated by hiatuses from the overlying zone of *E. huxleyi* and the underlying combined zone of *Discoaster brouweri/Discoaster pentaradiatus.* Along with both index species, the latter zone contains specimens of *D. tamalis,* probably reworked from the underlying combined zone of *Discoaster surculus/Discoaster tamalis.* As this species is not found below 485 cm, we assume that the core recovered more than 3.82 Ma (corresponding to the species FO, after Backman et al. 2012).

The correlation of nannofossil zones in the four cores studied from the Ioffe Drift (Fig. 11) demonstrates that they are better represented in core AI-3316, from northern slope of the drift than in the other three cores. In the drift summit cores

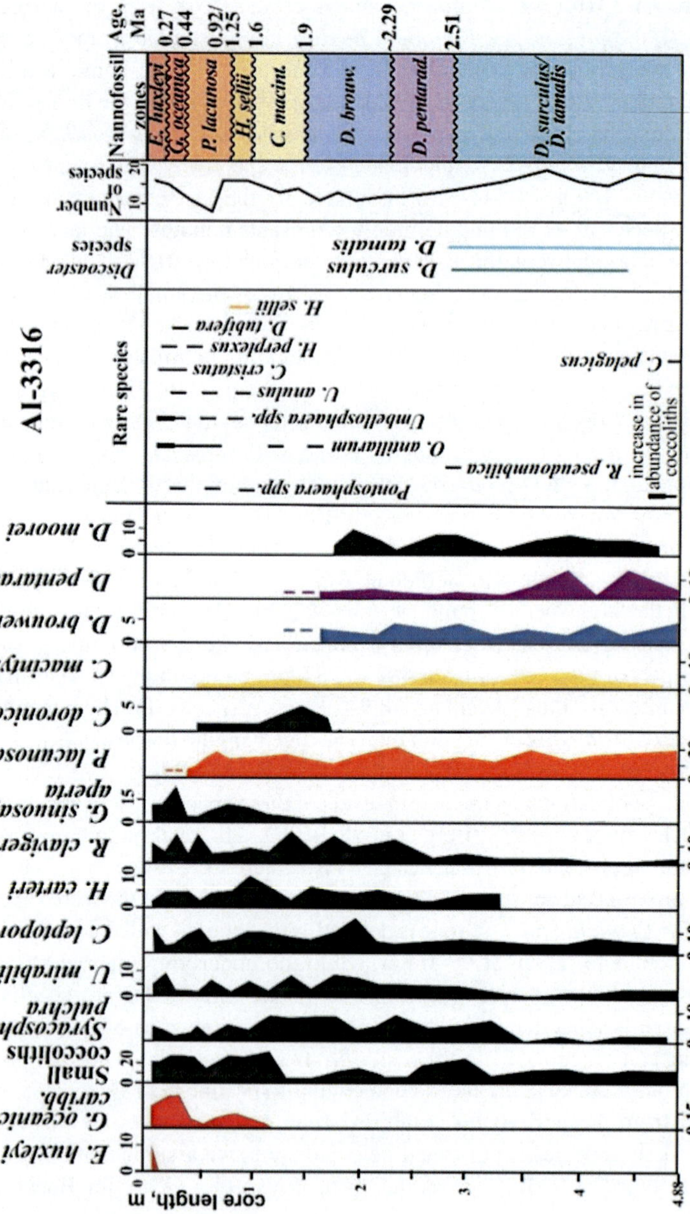

Fig. 9 Down-core distribution of species and nannofossil zones from core AI-3316. See text for the species full names and Fig. 1 for legend

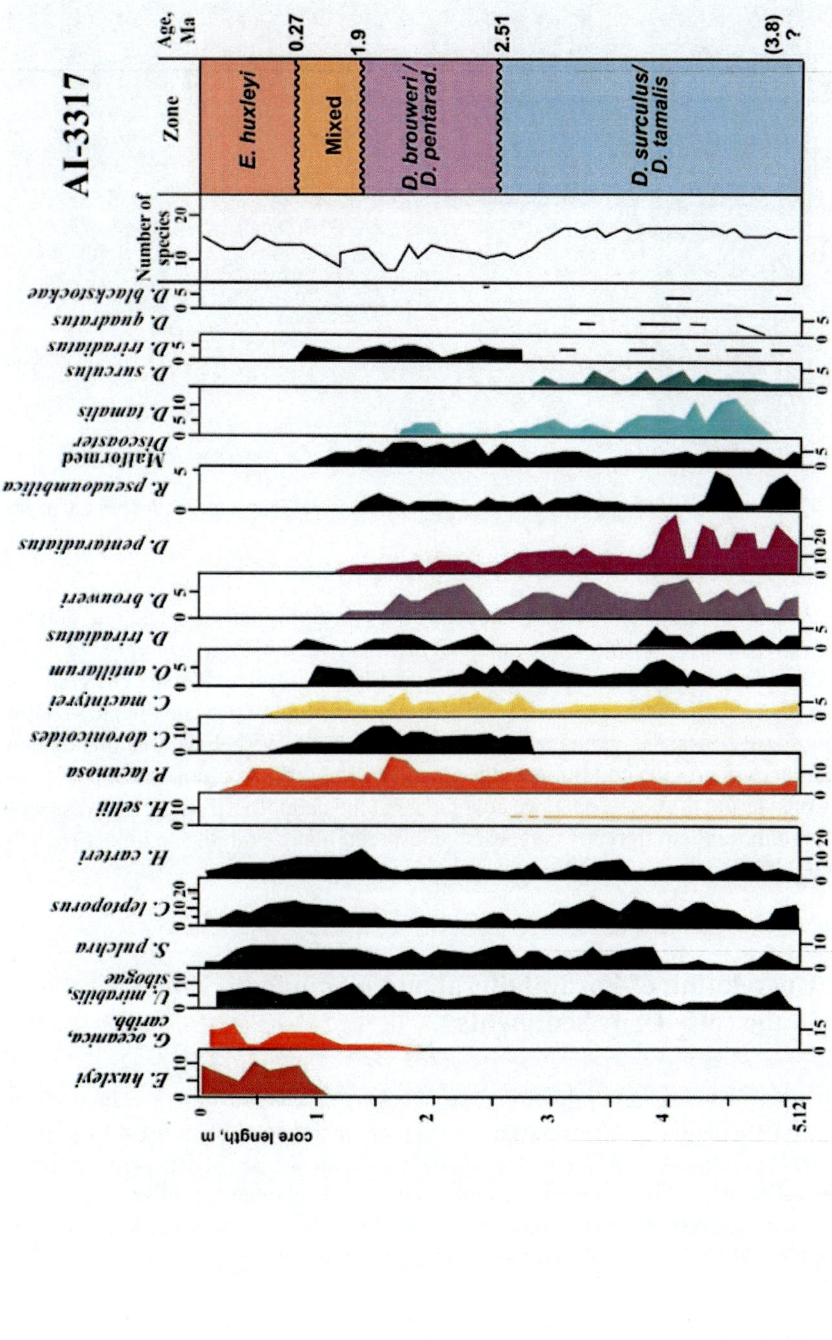

Fig. 10 Down-core distribution of species and nannofossil zones from core AI-3317. See text for the species full names and Fig. 1 for legend

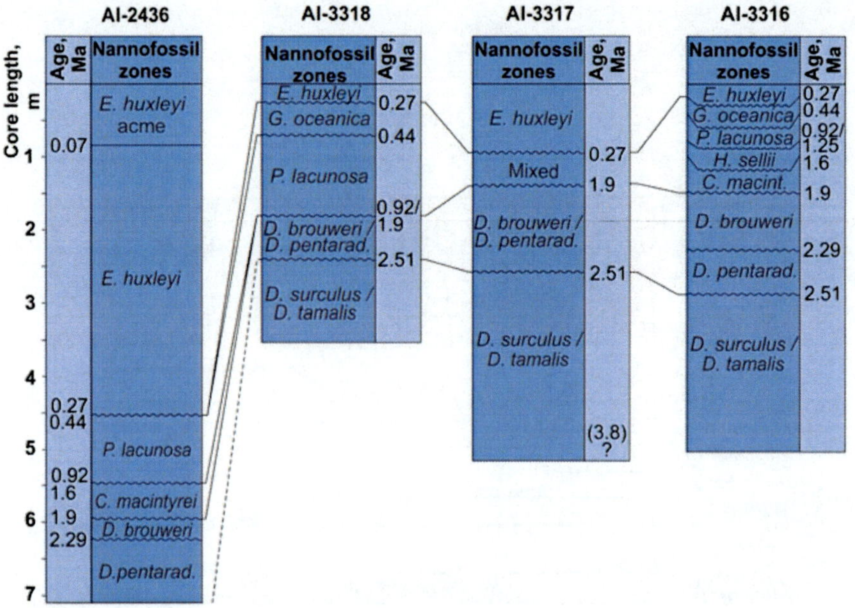

Fig. 11 Correlation of nannofossil zones in four cores studied. See text for the species full names and Fig. 1 for legend

(AI-2436 and AI-3318), some Pleistocene zones are commonly washed out, notably the *Gephyrocapsa oceanica*, the small *Gephyrocapsa*, the *Helicosphaera sellii* and the *Calcidiscus macintyrei* zones. In the Upper Pliocene–Lower Pleistocene sections, the *Discoaster* zones are commonly challenging to differentiate. The thickness of the nannofossil zones also varies with no sustainable trend. We believe that this reflects differences in erosion conditions rather than varying sedimentation rates.

Considering the above data, we conclude that the nannofossil stratigraphy, as well as foraminiferal stratigraphy, suggests widespread hiatuses in the sediment records of the Ioffe Drift area.

4 Correlation of Foraminiferal and Nannofossil Zones in the Ioffe Drift Sediments

In four cores where both microfossil groups were investigated a correlation of the foraminiferal and nannofossil zones seems reasonable, notably on the drift summit (cores AI-2436 and AI-3318) and northern slope (core AI-3316). The shifts in zonal boundaries, probably comprising hiatuses, can be explained by differences in the deposition and reworking of sand-size foraminiferal shells and tiny nannofossils (see Chaps. 6 and 10), and by the evolutional ages of species lineages that are documented by the FO and LO on the other hand. A good example is presented in Fig. 3 from core AI-3318, where the level of ~1.9 Ma is slightly higher in the nannofossil record than in the foraminiferal record. The longest (0.7–1 Ma) stratigraphic gap from this core is lower in the nannofossil record (0.92/1.9 Ma at ~180 cm, where three zones are absent) than in the foraminiferal record (0.81/1.47 Ma at ~130 cm, where the *Globorotalia crassaformis hessi* zone is absent).

In core AI-3317, with the strongest sediment and microfossil mixing (due to contamination, reworking and erosion), the boundary between foraminiferal zones of *Dentroglobigerina altispira* and *Globorotalia miocenica* at 3.13 Ma corresponds to the younger boundary of the combined *Discoaster tamalis/Discoaster surculus* and *Discoaster pentaradiatus/Discoaster brouweri* zones, at 2.51 Ma. This unexpected coincidence, along with the strong reworking of the fossils in the core section, confirms the occurrence of multiple erosional hiatuses (see also Chap. 9).

No significant increase in the thickness of the biostratigraphic zones or in the number of hiatuses was found between the northern slope (cores AI-3316 and AI-3317) and drift summit area (cores AI-2436, AI-3655 and AI-3318; Fig. 7). The same is true of core AI-3319, collected to the west-southwest of the Ioffe Drift.

5 Correlation of Biostratigraphic Zones from the Ioffe Drift and Rio Grande Rise

The foraminiferal and nannofossil zones in the current study correspond to those previously identified from the Pliocene–Pleistocene section of DSDP Site 516 (Barash et al. 1983; Barash 1988; Dmitrenko 1987); however, they are better represented in the latter section (Fig. 12). This is not surprising, as the foraminiferal preservation is excellent throughout the Pleistocene section of Site 516 (30°16.59′S 35°17.11′W), drilled on the summit of the Rio Grande Rise at a water depth of 1,313 m (Barash et al. 1983), but in the Ioffe Drift sections the preservation is variable.

Although contourite drifts are commonly characterized by high sedimentation rates (e.g., Robinson and McCave 1994; Stow et al. 2013; Rebesco et al. 2014), our previous (Ivanova et al. 2016, 2020) and current findings demonstrate a rather reduced thickness of all recovered upper Pliocene–Quaternary biostratigraphic zones in the sediment sections from the Ioffe Drift area, compared to DSDP Site 516 from the Rio Grande Rise (Ivanova et al. 2016, 2020; Fig. 12). For example, the FO of *Globorotalia truncatulinoides* (1.9 Ma) is identified at 980 cm below sea floor (bsf) at Site 516 (Barash et al. 1983) and at 600 cm in core AI-2436. The only exception, represented by the enhanced thickness of the uppermost zones of *G. ruber* pink and *E. huxleyi* from core AI-2436, resulted from the artificial stretching of the core during processing (Ivanova et al. 2016). Our cores AI-3320 and AI-3321 (see Table 1 in Chap. 4) from the Rio Grande Rise recovered only the uppermost foraminiferal zone of *Globigerinoides ruber* (pink), without any significant foraminiferal reworking. This is in line with the AMS-^{14}C dates suggesting a mean sedimentation rate of

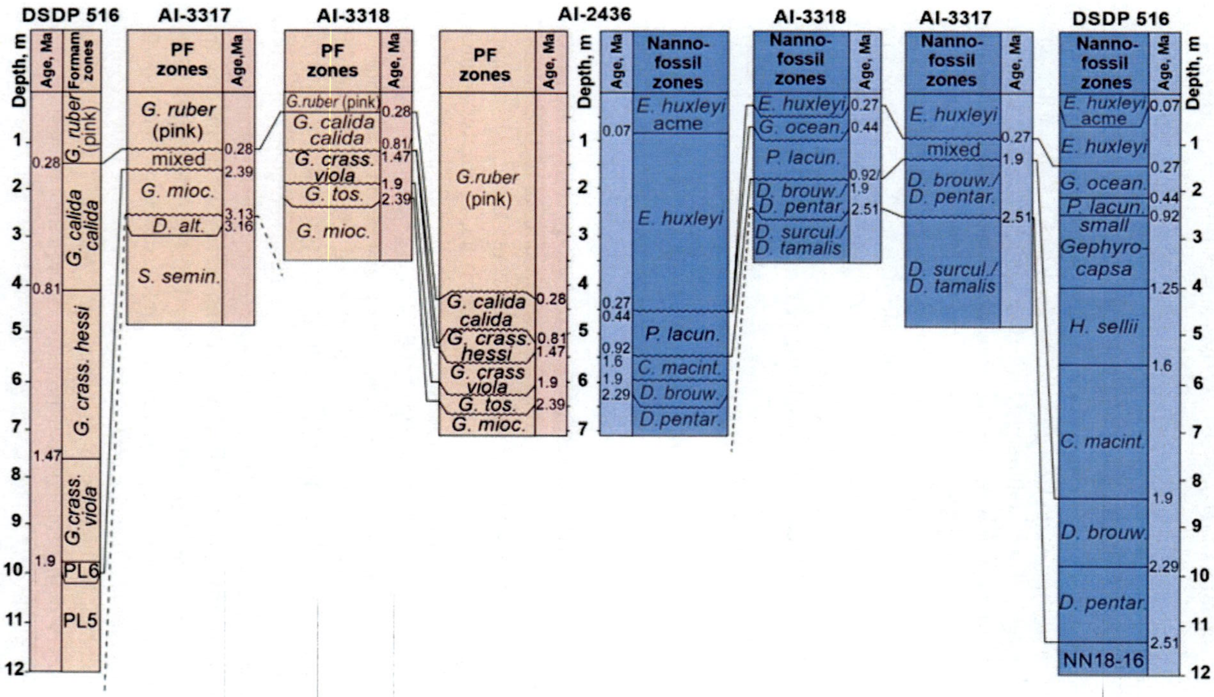

Fig. 12 Correlation of biostratigraphic zones in cores AI-2436 (Ivanova et al. 2016, 2020), AI-3318 (Ivanova et al. 2020), AI-3317, and DSDP Site 516. In the latter, foraminiferal zonation from Barash et al. (1983) and nannofossil zonation from Dmitrenko (1987) are used for the upper ~ 10 m, while the zonation of Berggren et al. (1983a) is applied below (using the zonal boundary ages from Wade et al. (2011) and Bergen et al. (2019). PF zones mean planktic foraminiferal zones. Assumed boundaries between zones and hiatuses are marked by dashed and undulated lines, respectively. See text for the species full names and Fig. 1 for legend

about 0.8 cm/ka for the upper part of both cores AI-3320 and AI-3321 (Table 1). Therefore, these cores support the finding of a greater thickness of biostratigraphic zones at the Rio Grande Rise than at the Ioffe Drift.

The difference in zone thickness in two neighboring areas can be explained first of all in terms of the dominant mechanism of sedimentation. According to Barker et al. (1983), on the Rio Grande Rise, pelagic sedimentation seems to prevail over lateral advection and other mechanisms. As shown in Chap. 6, in the Ioffe Drift the prevailing mechanism of sedimentation comprises lateral advection of biogenic material (originating from pelagic settling) by contour bottom currents. Although pelagic sedimentation is known to be quite slow in oligotrophic areas, the erosional effect of bottom currents in contourites, resulting in hiatuses and stratigraphic gaps, generally overtakes the effect of even relatively fast lateral sedimentation.

Furthermore, we argue that the occurrence of numerous hiatuses explains the aforementioned reduction in the thickness of biostratigraphic zones in the Ioffe Drift area (Ivanova et al. 2020; see also Chap. 9). Any significant impact of selective dissolution on foraminiferal fauna and on the thickness of biostratigraphic zones from the Ioffe Drift cores seems unlikely, apart from in core AI-3319 beyond the drift, from a water depth 4,066 m, and perhaps the lower part of core AI-3317. The aforementioned assumption follows from the generally good, rather good or even perfect preservation of foraminiferal tests in five cores, despite our core sites being much deeper (3,798–3,898 m) than those at DSDP Site 516 (1,313 m). A weak dissolution influence fits well with the established modern 4,050 m-depth of the foraminiferal lysocline in the area (Melguen and Thiede 1974) and probably indicates that during the Late Pliocene–Quaternary the boundary between Lower Circumpolar Deep Water (LCDW) and more corrosive Weddell Sea Deep Water (WSDW) was generally below 3,800 m. The only exception could be some relatively short-term intervals, like during MIS 3–2 for the Vema Channel area, as shown in Ovsepyan and Ivanova (2019).

If the identified hiatuses are of erosional origin, it means that surface bioproductivity and pelagic sedimentation in the Ioffe Drift area were not necessarily slower than at similar depths in the subtropical South Atlantic, where the thickness of Upper Pliocene–Quaternary sections is commonly greater than in our cores (Fig. 12; see also Chap. 9). Rather, the sediments were washed out by post-depositional erosion.

6 Conclusions

The sediment cores collected from the Ioffe Drift area recovered Upper Pliocene–Quaternary sediments. The biostratigraphic study of six sediment cores from the area permitted the identification of the eight foraminiferal and nine nannofossil (in four cores) zones described in Ivanova et al. (2016, 2020) and in Chap. 3. Core AI-3317, from the northern slope, spanned presumably the last ~4 Ma; that is, it penetrated deeper into the Upper Pliocene sediments than the other five cores.

The suggested foraminiferal and nannoplankton zonations collectively provide a robust stratigraphic framework for a multi-proxy study of the Ioffe Drift sediments and for paleoceanographic inferences.

The foraminiferal stratigraphic zonation developed for the six cores and visual estimates of the shells preservation collectively suggest that the drift slopes, unlike the summit, were affected by selective dissolution, notably in the late Pliocene. The strongest dissolution impact on foraminiferal fauna is documented from the lower half (i.e., in the Lower Pleistocene section) of core AI-3319, collected to the west-southwest of the Ioffe Drift.

The thickness of all biostratigraphic zones in the Ioffe Drift sections is significantly less than those in DSDP Site 516 on the nearby Rio Grande Rise. Nannofossil zones, notably those of *Gephyrocapsa oceanica, Helicosphaera sellii, Calcidiscus macintyrei, Pseudoemiliania lacunosa,* and especially of small *Gephyrocapsa*, are commonly washed out, like the foraminiferal zone of *Globorotalia crassaformis hessi*. In core AI-3317, the latter is mixed with the *Globigerinella calida calida, Globorotalia crassaformis viola* and *Globorotalia tosaensis* zones.

The reduced thickness and/or absence of foraminiferal and nannofossil zones, resulting in stratigraphic gaps in the sediments from the Ioffe Drift area, suggests the occurrence of several erosion hiatuses, as discussed in Chap. 9.

References

Backman J, Raffi I, Rio D et al (2012) Biozonation and biochronology of Miocene through Pleistocene calcareous nannofossils from low and middle latitudes. Newsl Stratigr 45:221–244. https://doi.org/10.1127/0078-0421/2012/0022

Barash MS, Oskina NS, Bylum NS (1983) Quaternary biostratigraphy and surface paleotemperatures based on planktonic foraminifers. In: Barker PF, Johnson DA, Carlson RL (eds) Initial reports of the deep-sea drilling project, 72. Government Printing Office, Washington, U.S, pp 849–869

Barash MS (1988) Quaternary paleooceanology of the Atlantic Ocean. Nauka (in Russian), Moscow

Barker PF, Johnson DA, Carlson RL et al (1983) Site 516: Rio Grande Rise. In: Barker PF, Johnson DA, Carlson RL (eds) Initial Reports of the deep-sea drilling project, 72. Government Printing Office, Washington, U.S, pp 155–338

Bergen JA, Truax S, de Kaenel E et al (2019) BP Gulf of Mexico Neogene astronomically-tuned time scale (BP GNATTS). Bull Geol Soc Am 131:1871–1888. https://doi.org/10.1130/B35062.1

Berggren WA, Aubry MP, Hamilton N (1983) Neogene magnetobiostratigraphy of deep-sea drilling project, site 516: Rio Grande Rise, South Atlantic. In: Barker PF, Carlson RL, Johnson DA (eds) Initial reports of deep-sea drilling project 72. Government Printing Office, Washington, U.S, pp 675–713

Berggren WA, Hamilton N, Johnson DA, Pujol C, Weiss W, Cepek P, Gombos AM Jr (1983) Magnetobiostratigraphy of deep-sea drilling project 72, sites 515–518: Rio Grande Rise (South Atlantic). In: Barker PF, Carlson RL, Johnson DA (eds) Initial reports of deep-sea drilling project 72. Government Printing Office, Washington, U.S, pp 939–948

Berggren WA, Hilgen FJ, Langereis CG et al (1995) Late Neogene chronology: new perspectives in high-resolution stratigraphy. Geol Soc Am Bull 107:1272–1287. https://doi.org/10.1130/0016-7606(1995)107%3c1272:LNCNPI%3e2.3.CO;2

Blow WH (1969) Late Middle Eocene to recent planktonic foraminiferal biostratigraphy. In: Brönnimann P, Renz HH (eds) Proceedings of the first international conference on planktonic microfossils, 1st edn. EJ Brill, Leiden, Geneva, pp 199–422

Bolli HM, Saunders JB (1985) Oligocene to Holocene low latitude planktonic foraminifera. In: Plankton Stratigraphy. Cambridge University Press, pp 155–262

Dmitrenko OB (1987) A detailed zonal scale of the quaternary bottom deposits based on coccoliths (on the Rio Grande rise in the Atlantic ocean). Oceanology 27(3):460–464 (in Russian with English translation)

Gartner S (1977) Calcareous nannofossil stratigraphy and revised zonation of the Pleistocene. Mar Micropal 2:1–25

Ivanova E, Murdmaa I, Borisov D et al (2016) Late Pliocene-Pleistocene stratigraphy and history of formation of the Ioffe calcareous contourite drift, Western South Atlantic. Mar Geol 372:17–30. https://doi.org/10.1016/j.margeo.2015.12.002

Ivanova E, Borisov D, Dmitrenko O, Murdmaa I (2020) Hiatuses in the late Pliocene-Pleistocene stratigraphy of the Ioffe calcareous contourite drift, western South Atlantic. Mar Pet Geol 111:624–637. https://doi.org/10.1016/j.marpetgeo.2019.08.031

Kennett JP, Srinivasan MS (1983) Neogene Planktonic Foraminifera: a phylogenetic atlas. Hutchinson Ross Publishing Company, Stroudsburg, Pensilvania

Melguen M, Thiede J (1974) Facies distribution and dissolution depths of surface sediment components from the Vema channel and the Rio Grande Rise (southwest Atlantic Ocean). Mar Geol 17:341–353. https://doi.org/10.1016/0025-3227(74)90096-6

Ovsepyan EA, Ivanov EV (2019) Glacial–interglacial interplay of southern- and northern-origin deep waters in the São Paulo Plateau-Vema channel area of the western South Atlantic. Palaeogeogr Palaeoclimatol Palaeoecol 514:349–360. https://doi.org/10.1016/j.palaeo.2018.10.031

Rahman A, Roth PH (1989) Late Neogene calcareous nannofossil biostratigraphy of the Gulf of Aden region. Mar Micropaleontol 15:1–27. https://doi.org/10.1016/0377-8398(89)90002-9

Rebesco M, Hernández-Molina FJ, Van Rooij D, Wåhlin A (2014) Contourites and associated sediments controlled by deep-water circulation processes: state-of-the-art and future considerations. Mar Geol 352:111–154. https://doi.org/10.1016/j.margeo.2014.03.011

Robinson SG, McCave IN (1994) Orbital forcing of bottom-current enhanced sedimentation on Feni drift, NE Atlantic, during the mid-Pleistocene. Paleoceanography 9:943–972. https://doi.org/10.1029/94PA01439

Schiebel R, Hemleben C (2017) Planktic foraminifers in the modern ocean. Springer, Berlin

Stow DAV, Hernandez-Molina FJ, Alvarez-Zarikian CA (2013) Expedition 339 Scientists (2013) Expedition 339 summary. In: Stow DAV, Hernández-Molina FJ, Alvarez Zarikian CA, Expedition 339 Scientists (eds) Proceedings of the IODP, 339. Integrated Ocean Drilling Program Management International, Inc., Tokyo. https://doi.org/10.2204/iodp.proc.339.101.2013

Wade BS, Pearson PN, Berggren WA, Pälike H (2011) Review and revision of Cenozoic tropical planktonic foraminiferal biostratigraphy and calibration to the geomagnetic polarity and astronomical time scale. Earth-Science Rev 104:111–142. https://doi.org/10.1016/j.earscirev.2010.09.003

Ferromanganese Nodules

Natalia Shulga, Ivar Murdmaa, Olga Dara, and Konstantin Ryazantsev

Abstract Ferromanganese nodules found in the upper 4 m of core AI-2436 (which is retrieved from the Ioffe Drift summit) were subjected to complex study. This included description of distribution, morphology, internal microstructure, investigation of chemical and mineral compositions, in order to understand their genesis. Three well-developed zones were distinguished within the nodules: the outer crust, the transitional and central zones. Geochemical data indicate a hydrogenetic origin for the nodules, with an average ratio Mn/Fe of 0.84. A lack of dissolution evidence (visible and chemical) suggests the domination of oxic conditions in the calcareous ooze of the core accommodating the nodules. The Ce/La ratio has a value characteristic of nodules swept by the Lower Circumpolar Deep Water (LCDW), which is considered to be responsible for the Ioffe Drift's formation. The specific shape (cylindrical with wavy surface, rounded triangle) suggests formation of nodules after the reworking of bottom surface sediments by benthic fauna. Poor crystallinity and absence of crystalline Fe-rich phases possibly reflects early burial of the relatively young nodules (with the geological age not older than 0.26 Ma).

N. Shulga (✉) · I. Murdmaa · O. Dara
Shirshov Institute of Oceanology, Russian Academy of Sciences, Moscow, Russia

K. Ryazantsev
Vernadsky Institute of Geochemistry and Analytical Chemistry, Russian Academy of Sciences, Moscow, Russia
e-mail: ryazan-konst@rambler.ru

I. Murdmaa and E. Ivanova (eds.), *The Ioffe Drift*, Springer Geology,
https://doi.org/10.1007/978-3-030-82871-4_8

1 Introduction

Ferromanganese (Fe–Mn) nodules, as polymetallic ore deposits, have gained recognition not only for their high iron and manganese content but also for the presence of critical high-technology metals such as copper, nickel, cobalt, zinc, molybdenum, rare earth elements and yttrium (REY) (Halbach et al. 1988; Hein and Koschinsky 2014). These deposits are formed by: (a) hydrogenetic precipitation or the accumulation of colloidal metal oxides that precipitate directly from cold ambient ocean water; (b) oxic/suboxic diagenesis from sediment-pore fluids; and (c) precipitation of Fe oxyhydroxide and Mn oxide directly from hydrothermal solutions in areas such as mid-ocean ridges, back-arc basins and hotspot volcanoes (Dymond et al. 1984; Baturin 1986; Glasby 2006; Hein and Koschinsky 2014; Hein et al. 2015; Kuhn et al. 2017; Josso et al. 2017; Reykhard and Shulga 2019). Slow growth is characteristic of hydrogenetic nodules (about 2 mm/Ma) and Mn/Fe ratios <2.5, while diagenetic nodules grow faster (about 10–100 mm/Ma) and have Mn/Fe ratios of >5 (Skornyakova and Murdmaa 1992; Hlawatsch et al. 2002; Glasby 2006; González et al. 2012; Menendez et al. 2019). The growth rate of hydrothermal deposits is generally extremely high (>1,000 mm/Ma) (Glasby 2006). Fe–Mn nodules are found throughout the World Oceans. However, wide dense nodule ore fields occur only in pelagic areas with very low sedimentation rates mainly in Pacific and Eastern Indian oceans. In the Atlantic, ferromanganese nodules are mainly represented by the hydrogenetic variety (Skornyakova and Murdmaa 1995). Their distribution is also related to the nature of the oceanic bottom-water flow (Kasten et al. 1998; Glasby 2006).

Ferromanganese nodules were found in several cores from the Ioffe Drift (AI-2436, AI-3316, AI-3318) and in grab samples (AI-3653, AI-3654) from the fault channel north of the drift. Buried Fe–Mn nodules in various depth intervals from core AI-2436 were analyzed to investigate their formation, morphology, geochemical and mineralogical composition. Buried nodules are less known and studied than surface nodules.

2 Methods

Numerous Fe–Mn nodules found in sediment core AI-2436 were subjected to detailed analysis (Fig. 1 in Chap. 1). For a lithological description of core AI-2436, see Chap. 6. Nodules of varying sizes and morphology from the different core intervals were studied to ascertain their chemical and mineralogical compositions using inductively coupled plasma mass and optical emission spectrometry (ICP-MS, ICP-OES), X-ray diffraction (XRD) and scanning electron microscope Energy Dispersive X-Ray Spectroscopy (SEM–EDS). Structural and textural features, as well as the chemical composition and mineralogy, were analyzed in bulk samples and in subsamples of air-dried nodules, separately for the nodules' crust and central zone. Separation was performed on the dried samples manually.

2.1 Geochemistry

The bulk nodules (25 samples), their outer crusts (3 samples) and central zones (3 samples), were tested for the amount of major and some trace elements (Na, Mg, P, S, K, Ca, Al, Ti, V, Cr, Mn, Fe, Co, Ni, Cu, Zn and Sr) by ICP atomic emission spectrometry (ICP-AES) (ICAP-61, Thermo Jarrell Ash, USA). The trace-element (Li, Sc, Cr, Ni, Cu, Zn, Sr, Mo, Cd, Ba, W and Pb) and REY concentrations in the samples listed above were determined by ICP-MS (X-7, Thermo Elemental, USA). Samples were dried, crushed to powder size and dissolved with $HNO_3 + H_2O_2$ (4:1 by volume, Merck) in an autoclave system. A detailed description of the methodology and accuracy is given in Karandashev et al. (2008).

2.2 Scanning Electron Microscopy (SEM–EDS)

Investigations of the nodule microstructures were made by SEM analyses with an LEO Supra 50 VP instrument (Carl Zeiss, Germany) Tescan scanning electron microscope Mira 3 with analytical equipment of the Oxford Instruments AztecLive Automate with detector Max 80. Polished thin sections of 150 μm thickness were prepared. Fragments of the outer crust and central zone of buried nodules from various intervals were selected for analysis. The surface of the samples was covered with 15-nm-thick gold with Balzers SCD 030 equipment. Major element concentrations were determined using energy-dispersive X-ray spectroscopy on an X-MAX 80 (EDS, Oxford Instruments, UK). Analyses were conducted at 20 kV using a diaphragm of 60 μm. Data processing was carried out using the INCA Oxford software package.

2.3 XRD Mineralogy

The mineralogy of the outer crust, transition and central zones of three typical Fe–Mn nodules were analyzed by XRD using a D8 ADVANCE (Bruker AXC) diffractometer, Cu-Kα radiation (40 kV and 40 mA) equipped with a LYNXEYE detector. To determine the phase composition of nodules and identify metastable water-containing phases for each sample, two X-ray diffraction patterns were carried out according to the technique described by Chukhrov et al. (1976) and Smetannikova (1988). First, the nodules were scanned by XRD from 2.5° to 70° (2θ) with a step size of 0.02° (2θ). Second, samples were heated to 105 °C for one hour and scanned from 2.5° to 14° (2θ) with a step size of 0.02°. The DIFFRAC.EVA program was used to interpret the spectra and calculate the diffraction patterns. A semi-quantitative assessment of the mineral phases in the samples was performed using corundum numbers from the PDF-2 ICDD database.

3 Distribution, Structure, and Composition of Nodules from the Core AI-2436

3.1 Distribution and Morphology

The buried ferromanganese nodules (altogether 25) were found in core AI-2436 in the interval from 15 to 416 cm below the observed coretop after extraction of the deformed core from the corer. The greatest number of nodules is found in the upper core interval from 15 to 82 cm. The middle interval, 82–217 cm, contained less abundant nodules and only rare ones occurred in the lower interval, 217–416 cm.

The shape of nodules is the most variable in the upper nodule bearing layer (15–82 cm), where along with common irregular spheroidal nodules 1.5–3 cm in diameter (Plate 1, type 1) some nodules resemble grown together of three spherules as a triangle with rounded angles (Plate 1, type 2). Their size varies within 2.5–2.7 cm. Specific cylindrical forms up to 6 cm long occur visually resembling grown together 2–5 spherules in straight lines (Plate 1, type 3).

3.2 Microstructures and Mineralogy

On the basis of the SEM examination, three zones were distinguished in the structure of all types of nodules: an outer crust; a transitional zone; and a central zone (Plate 2). XRD data show poor crystallinity in all the nodules analyzed.

The outer crust is thin, black with white calcite powder (from the accommodating sediments) and dense but fragile. It easily detaches from the body of the nodule. The outer part of the crust has a microglobular structure (Plate 3a). Clusters of white particles comprising biogenic mineral components (coccoliths and biogenic detritus), are observed in the space between the microglobules (Plate 3f). The SEM–EDS examination found that the microglobules contain Mn (27.4%), Fe (22.2%), Ti (1.40%) and Ni (0.37%), on average. The mineral composition of the outer crust is dominated by vernadite with impurities from an admixture of quartz and calcite (Table 1). The Backscattered-Electron (BSE) image shows a columnar growth pattern and dendritic structure typical of ferromanganese crusts (Plate 2).

The transitional zone is dense. It contains alternating micro-layers of various colors (from gray, dark gray to black) and has a thin micro-layered structure. Buserite and asbolan dominate in the mineral composition. Traces of birnessite and vernadite are also detected, as well as an admixture of quartz and feldspar (Plates 1 and 2, Table 1).

The central zone is composed of a solid, dense substance, irregular in color. Powders of light brown matter of unknown composition (possibly clay) and biogenic particles from adjacent sediment were noted. XRD revealed that among the minerals dominate buserite, asbolan, and birnessite, with an admixture of quartz and feldspars (Table 1). A multiple thin-layering and colloform intergrowth patterns, as well as a

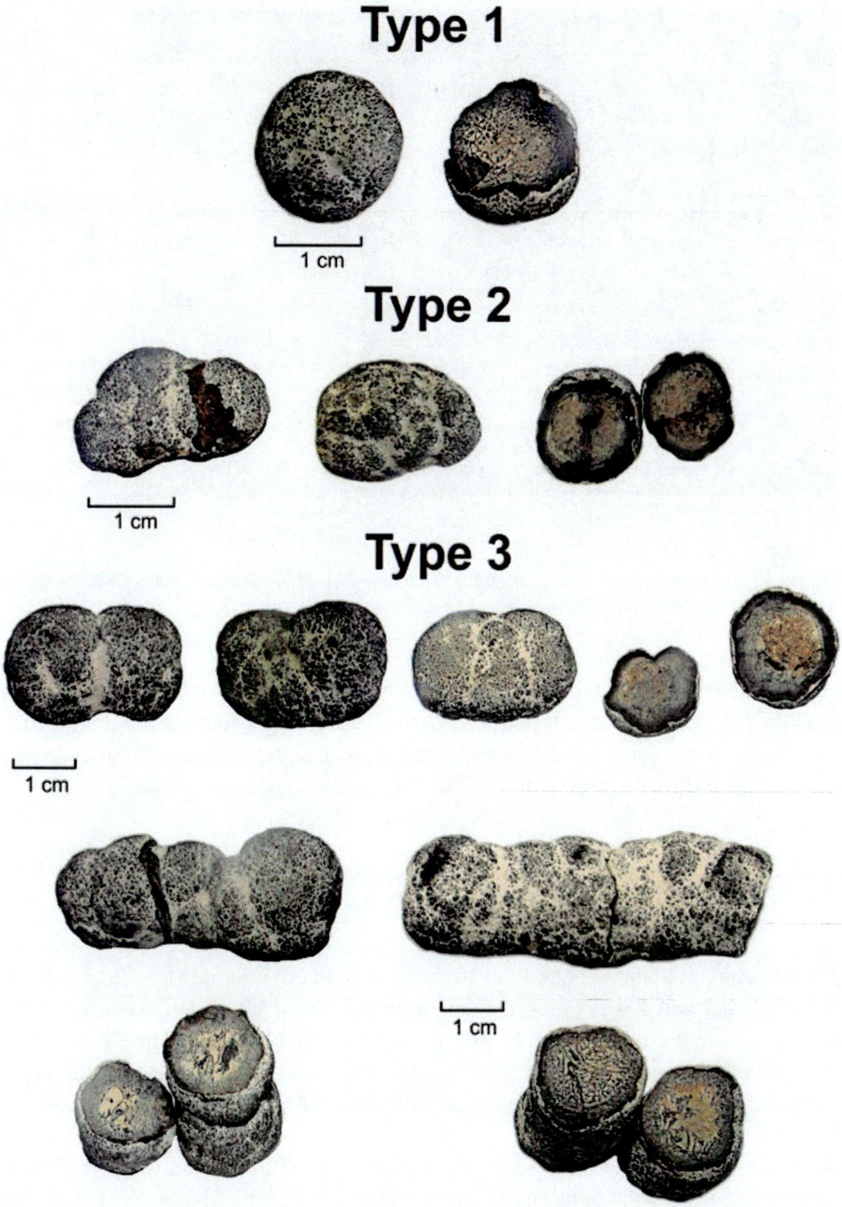

Plate 1 Typical buried nodules collected from core AI-2436

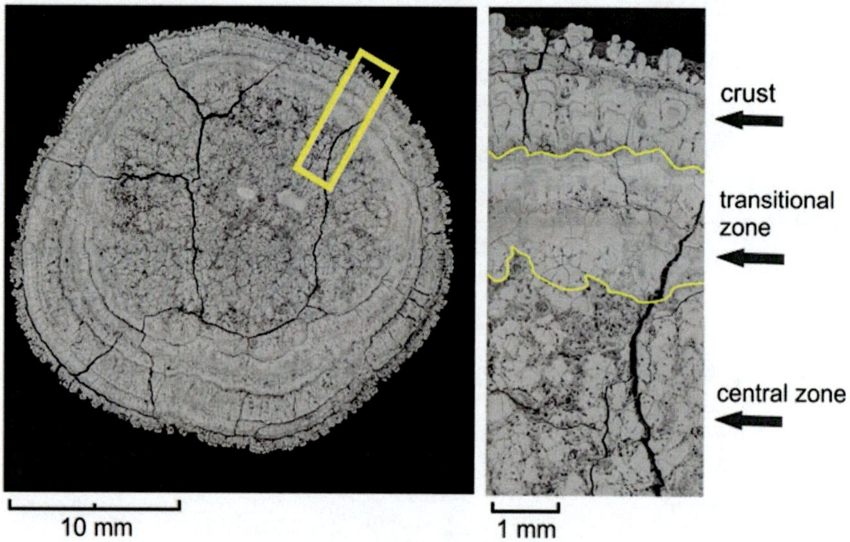

Plate 2 Thin section (BSE image) of nodule from core AI-2436. Yellow rectangle shows the area with detailed textures and structures image in different nodule zones

radial structure were found in the cross-section of the central zone (Plate 2). Plenty of randomly distributed cavities and well-distinguished aggregates of biogenic detritus are present in this zone. The SEM–EDS images reveal the presence of non-crystalline Fe and Mn authigenic segregations (Plate 3c, d, e).

3.3 *Geochemistry*

The Mn content in the bulk nodules from the upper core interval (15–82 cm), middle (82–217 cm) and lower (217–416 cm) ranges from 13.9 to 19.7wt%, from 14.4 to 17.5wt% and from 14.0 to 15.2wt%, respectively (Table 2). The Fe content of the bulk buried nodules within all intervals is 18.8wt% on average, with no significant differences. The Mn/Fe ratios are similar and range from 0.71 to 0.96. The Ni content ranges from 0.28 to 0.30wt%. The Cu content shows no considerable difference between the analyzed samples and is of 0.14wt% on average. The Co content increases with depth: in the upper layer (15–82 cm) the nodules contain less Co (0.38wt%) than those in the lower core intervals (0.41wt%). Thus, the nodules are characterized by low content of three major ore elements, Ni, Cu, Co, which is common for the hydrogenetic ferromanganese nodules. A positive cerium anomaly ($Ce_{an} = 3.2$, on average) and a negative yttrium anomaly ($Y_{an} = 0.75$) were revealed. The Ce/La ratio reaches high values, from 7.0 to 8.0, with an average of 7.3.

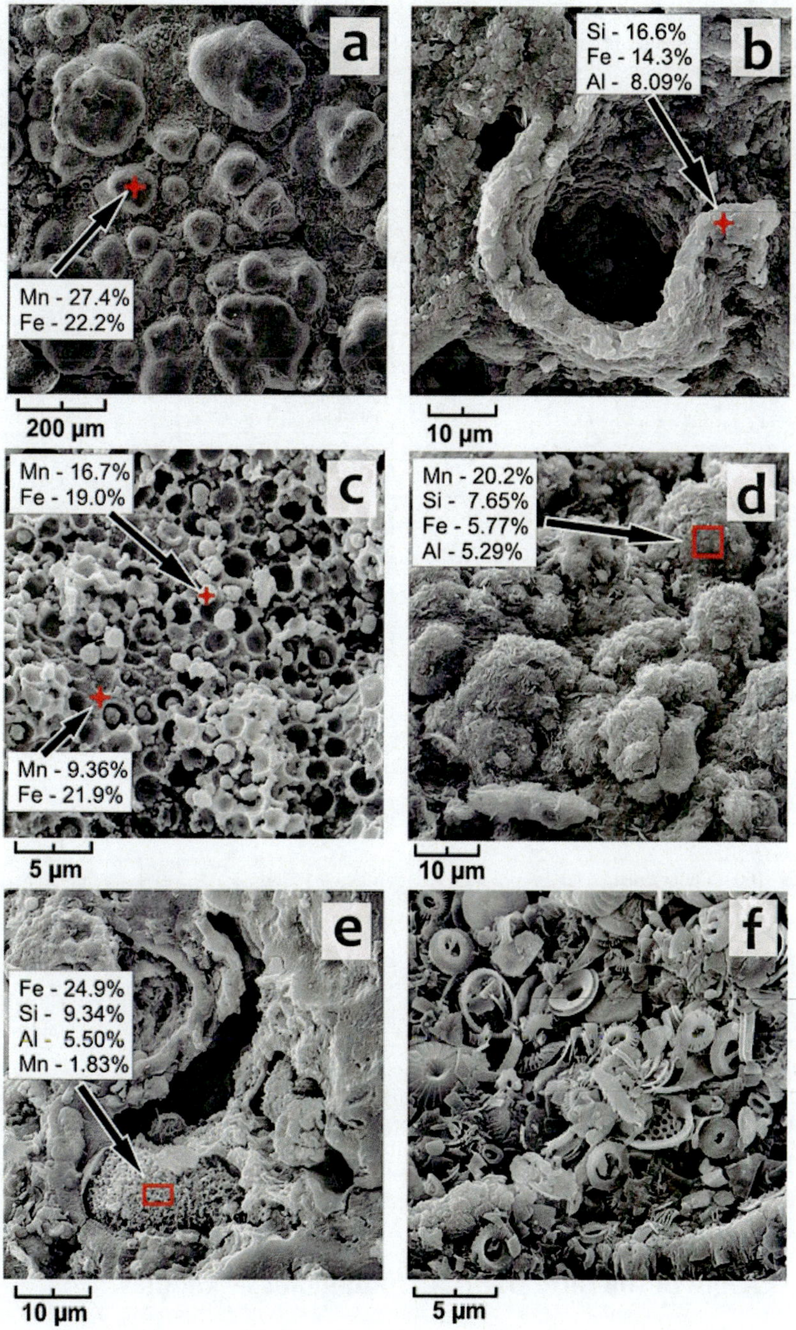
a
Mn - 27.4%
Fe - 22.2%
200 μm
b
Si - 16.6%
Fe - 14.3%
Al - 8.09%
10 μm
c
Mn - 16.7%
Fe - 19.0%
Mn - 9.36%
Fe - 21.9%
5 μm
d
Mn - 20.2%
Si - 7.65%
Fe - 5.77%
Al - 5.29%
10 μm
e
Fe - 24.9%
Si - 9.34%
Al - 5.50%
Mn - 1.83%
10 μm
f
5 μm

◀**Plate 3** SEM images of microstructures in the buried Fe–Mn nodules from core AI-2436 core: **a** globular microtexture of the outer crust; **b** authigenic Si-rich substance; **c** cellular-globular Fe-rich structure of the central zone; **d** microglobular fibrous Mn-rich structure of the central zone; **e** Fe-rich structure of the central zone; and **f** clusters of coccoliths and biogenic detritus in the interglobular area of the outer crust

Table 1 Mineral composition of selected buried Fe–Mn nodules from core AI-2436. Samples prepared from spheroidal (20–21 cm), irregular nodular (123–124 cm) and polynucleate cylindrical (worm-like) (158–160 cm) nodules

Interval, cm	Sample	Mineral					Impurities (<8%)
		10 Å	10 Å	10 Å	7 Å	δ-MnO_2	
		Buserite I[1]	Unstable buserite[2]	Buserite, asbolan[3]	Birnessite	Vernadite	
20–21	Outer crust	nd*	nd	+	+	+++	Quartz, calcite
	Transitional zone	+	++	++	nd	nd	Quartz, calcite, feldspar
	Central zone	nd	++	+++	+	nd	Quartz, calcite, feldspar
123–124	Outer crust	+	++	nd	nd	+	Quartz, calcite
	Transitional zone	nd	++	nd	nd	nd	Quartz, feldspar
	Central zone	nd	nd	++	++	nd	Quartz, feldspar
158–160	Outer crust	nd	nd	nd	nd	+++	Quartz, calcite
	Transitional zone	nd	++	++	+	+	Quartz, feldspar
	Central zone	nd	nd	+++	nd	nd	Quartz, feldspar

*nd—not detected, +—traces, ++—notable amount, +++—abundant
[1]Manganese hydroxide, the basal reflex 001 after heating at 105 °C shifts to the 7 Å region of birnessite
[2]Manganese hydroxide, the basal reflex 001 shift toward larger angles when heated up to 105 °C
[3]A group of manganese hydroxides (such as asbolan, buserite-II, etc.), in which the basal reflex 001 remains at the same position after heating to 105 °C

4 Origin of the Ioffe Drift Ferromanganese Nodules

Fe–Mn nodules lying on the seafloor's surface are of varying compositions, depending on the depositional environment, the content of organic matter in the

Table 2 Major, trace and REE concentrations in the buried Fe–Mn nodules at different intervals of core AI-2436. Intervals represent buried nodules: upper (15–82 cm), middle (82–217 cm) and lower (217–416 cm)

Interval	15–82 cm			82–217 cm			217–416 cm		
Element	Min	Max	Average	Min	Max	Average	Min	Max	Average
Fe %	14.5	21.0	18.8	16.3	20.7	19.1	16.4	19.2	17.9
Mn	13.9	19.7	15.8	14.4	17.5	16.1	14.0	15.2	14.6
Al	2.28	4.11	3.09	2.46	3.99	3.18	2.38	3.67	3.21
Mg	1.24	1.55	1.43	1.29	1.57	1.46	1.36	1.44	1.40
Ca	1.45	2.50	1.97	1.75	2.38	1.98	1.42	1.95	1.66
Na	1.18	1.37	1.23	1.16	1.39	1.23	1.03	1.30	1.19
K	0.37	0.77	0.51	0.38	0.78	0.52	0.35	0.70	0.56
Ti	0.74	1.14	1.00	0.80	1.11	1.01	0.79	1.02	0.91
P	0.34	0.44	0.41	0.36	0.44	0.41	0.39	0.42	0.40
S	0.21	0.25	0.24	0.23	0.25	0.24	0.20	0.23	0.22
As, ppm	173	289	243	192	278	250	197	264	231
Ba	847	1154	1015	938	1278	1206	844	964	914
Be	5.35	6.01	5.67	5.42	5.92	5.67	5.17	5.67	5.43
Bi	16.5	19.1	17.6	16.5	19.0	17.5	16.4	18.3	17.7
Cd	4.19	6.34	5.63	4.59	6.49	5.76	5.22	5.92	5.64
Co	2715	4441	3816	2983	4202	3880	3743	4795	4110
Cr	15.7	39.7	25.3	17.9	33.1	25.8	23.8	34.6	29.2
Cs	0.56	1.16	0.77	0.58	1.22	0.80	0.50	1.02	0.81
Cu	992	1758	1386	1121	1712	1452	1235	1319	1285
Ga	4.16	17.7	6.66	4.72	6.75	6.90	5.66	16.7	12.0
Hf	9.44	13.4	11.9	10.1	12.9	12.1	11.2	14.0	12.0
Li	22.3	54.8	40.1	27.2	54.0	42.8	36.9	45.5	41.7
Mo	202	260	231	214	243	235	233	257	240
Nb	64.2	95.2	83.9	67.1	93.0	85.6	72.4	98.9	83.3
Ni	2142	3276	2808	2292	3456	2887	2843	3085	2950
Pb	777	1274	1046	752	1196	1050	847	1238	993
Rb	7.04	18.6	11.4	7.31	19.0	11.8	7.52	18.0	13.4
Sb	35.5	43.0	39.3	37.8	43.5	40.9	36.7	41.4	39.2
Sc	10.7	17.1	13.4	11.2	15.7	13.5	13.4	16.9	14.9
Sn	1.39	4.08	2.62	0.44	4.19	2.73	3.80	4.58	4.10
Sr	808	1017	903	801	961	896	852	973	892
Ta	1.25	1.62	1.47	1.30	1.62	1.48	1.16	1.53	1.37
Te	31.8	40.0	36.2	33.8	40.3	37.5	33.2	40.3	36.1
Th	59.4	75.8	68.1	63.2	72.5	69.0	67.8	80.1	72.7

(continued)

Table 2 (continued)

Interval	15–82 cm			82–217 cm			217–416 cm		
Element	Min	Max	Average	Min	Max	Average	Min	Max	Average
Tl	93.6	184	139	93.3	155	141	137	179	153
U	8.65	9.62	9.14	8.55	9.67	9.05	8.89	9.80	9.32
V	471	710	603	474	616	594	547	708	592
W	36.1	128	54.0	40.8	58.0	56.3	48.2	57.6	51.2
Y	145	165	156	133	162	152	155	164	160
Zn	530	657	608	548	667	621	572	623	601
Zr	409	551	508	444	531	505	481	554	504
ΣREE	1938	2393	2175	2005	2292	2168	2120	2429	2241
Mn/Fe	0.71	1.36	0.85	0.71	0.96	0.88	0.75	0.87	0.82

sediments and nodules, the composition of the underlying sediments and the supply of elements from the pore water and overlying seawater (Kasten et al. 1998; Klevenz et al. 2010; Wegorzewski and Kuhn 2014; Wegorzewski et al. 2015). Being buried, nodules may either continue to grow, stop growing or dissolved (Heye and Marchig 1977; Pattan and Parthiban 2007). The processes taking place within buried nodules and adjacent sediments remain understudied. A lack of dissolution evidence (visible and chemical) in all the analyzed nodules suggests the domination of oxic conditions in the calcareous oozes throughout the core AI-2436, which is also confirmed by their yellowish color shades. Under such oxic conditions, the elements distribution patterns of surface, subsurface and deeply buried nodules are similar.

Fe–Mn deposits can be classified on the basis of their major chemical elements (Fe, Mn, Cu–Ni–Co), using the E. Bonatti ternary diagram (Bonatti et al. 1972). Bau et al. (2014) developed an approach to determine nodule origin using diagrams of the ratio of the rare earth elements. The samples studied fall into hydrogenetic (seawater-sourced) area, clearly separate from nodules of mixed and diagenetic origin (Fig. 1).

The Ce/La ratio in deep-water Fe–Mn nodules has been used in previous studies to track the Antarctic Bottom Water (AABW) flow path in the Southwest Pacific as an indicator of the redox potential of the environment (Glasby et al. 1987). Kunzendorf et al. (1993) showed that the highest values of the Ce/La ratio and the content of rare earth elements in nodules are observed in the central part of the Pacific Ocean, where the AABW flux controls bottom circulation. In the South Atlantic, a similar relationship has been traced between the Ce/La ratio in nodules and water masses affecting the bottom in the nodule area (Kasten et al. 1998), and the data point to a more complex bottom-water circulation pattern in the Atlantic than in the Pacific. The authors explain the higher Ce/La ratios in the nodules from the Angola and Brazilian basins by the influence of the more oxygenated North Atlantic Deep Water (NADW). These data confirm the potential for assessing the contribution of various water masses to the formation of deep-water ferromanganese nodules. The Ce/La

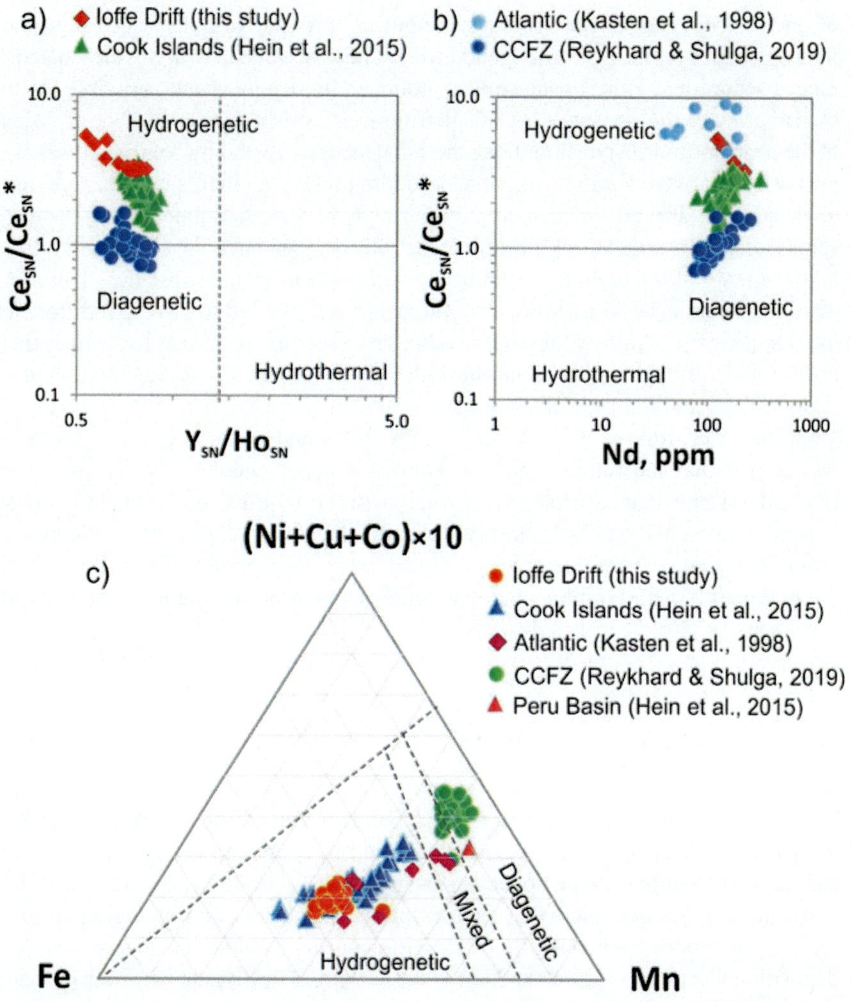

Fig. 1 Relationship between: **a** CeSN/CeSN* ratio versus Nd concentration and **b** CeSN/CeSN* ratio in bulk nodules of different genesis, where CeSN* = 0.5 × LaSN + 0.5 × PrSN and SN–shale-normalized; **c** ternary diagram of Fe–Mn-(Ni + Cu) × 10 according to Bonatti et al. (1972), showing the geochemical relationship between different genetic types of nodules (bulk samples)

ratio (7.3) in the studied nodules has a value corresponding to influence by the inflow of LCDW (Kasten et al. 1998). This confirms that all the buried nodules of the Ioffe Drift were formed in a well-oxygenated environment.

Specific shapes of nodules, especially in the upper layer of their occurrence, including straight cylindrical ones (Type 3) with wavy surfaces, as well as the "triangle" ones (Type 2), their almost equal size parameters, and absence of nuclei, need explanation. Hydrogenetic growth itself, well proven by geochemical data, could not create this rather complicated morphology if it was not predicted by the shape of nuclei. This suggests a hypothesis about the primary origin of such repeatedly figured shapes from the forms generated by benthic fauna, which were coated by mucus secretions. This substance was produced by benthic fauna, possibly feeding by gathering of the very poor organic detritus reaching the bottom surface or floating in the near-bottom suspension. If so, the substances extracted by benthic fauna might primarily be covered with ferromanganese film preserving their shape that continues to grow as a hydrogenetic crust until the burial by very slow sedimentation. Inside the semi-isolated interior of nodules, diagenetic processes have developed first driven by the energy of the organic matter decay within the biogenic substance. This might continue after burial of nodules. The internal diagenesis led to principal differences in ferromanganese mineralogy of the outer crust (vernadite typical for hydrogenetic crusts) and central zone (asbolan-buserite which dominates in diagenetic nodules (Uspenskaya and Skornyakova 1991)).

So far as crystalline Fe-rich minerals are not found in the nodules by the XRD, we assume that the rather high iron content is represented by X-ray amorphous (colloidal?) substance. Along with total poor crytallinity of authigenic Fe–Mn minerals noted above, this possibly indicates that the hydrogenetic nodules were still rather young, incompletely crystallized, when they buried. This is on line with the maximal biostratigraphic age of sediments which buried the nodules (0.27 Ma, see Chap. 7).

5 Conclusions

The composition of the buried nodules found in the sediments of core AI-2436 indicates a hydrogenetic origin and consistent ore formation conditions during the period corresponding to the upper 4 m of the core.

Inside the nodules, plenty of cavities partially filled by ferromanganese matter are well developed, indicating fluid circulation during the nodules' genesis and post-depositional activities. The formation of the nodules shape on the substrate produced during the reworking of sediments by benthic fauna is hypothesized. We also assume that internal diagenesis led to re-precipitation of ferromanganese minerals in the central zone of nodules and to a total replacement with ferromanganese minerals of the material of accommodating sediments reworked by benthic fauna.

The supply of the geochemical elements needed for nodule growth is mainly associated with the contour current of the Lower Circumpolar Deep Water (LCDW), which played a crucial role in the formation of the Ioffe Drift.

Poor crystallinity and absence of crystalline Fe-rich phases possibly reflects early burial of the relatively young nodules (with the geological age not older than 0.27 Ma).

References

Baturin GN (1986) Geochemistry of oceanic ferromanganese nodules. Nauka, Moscow

Bau M, Schmidt K, Koschinsky A et al (2014) Discriminating between different genetic types of marine ferro-manganese crusts and nodules based on rare earth elements and yttrium. Chem Geol 381:1–9. https://doi.org/10.1016/j.chemgeo.2014.05.004

Bonatti E, Kraemer T, Rydell H (1972) Ferromanganese deposits on the ocean floor

Chukhrov FV, Zvyagin BB, Yermilova LP, Gorshkov AI (1976) Mineralogical criteria in the origin of marine iron-manganese nodules. Miner Depos 11:24–32. https://doi.org/10.1007/BF00203092

Dymond J, Lyle M, Finney B et al (1984) Ferromanganese nodules from MANOP Sites H, S, and R—Control of mineralogical and chemical composition by multiple accretionary processes. Geochim Cosmochim Acta 48:931–949. https://doi.org/10.1016/0016-7037(84)90186-8

Glasby GP (2006) Manganese: predominant role of nodules and crusts. In: Shulz HD, Zabel M (eds) Marine geochemistry. Springer, pp 371–427

Glasby GP, Gwozdz R, Kunzendorf H et al (1987) The distribution of rare earth and minor elements in manganese nodules and sediments from the equatorial and S.W. Pacific Lithos 20:97–113. https://doi.org/10.1016/0024-4937(87)90001-6

González FJ, Somoza L, León R et al (2012) Ferromanganese nodules and micro-hardgrounds associated with the Cadiz contourite channel (NE Atlantic): palaeoenvironmental records of fluid venting and bottom currents. Chem Geol 310–311:56–78. https://doi.org/10.1016/j.chemgeo.2012.03.030

Halbach P, Friedrich G, von Stackelberg U (1988) The manganese nodule belt of the Pacific Ocean: geological environment, nodule formation, and mining aspects. Enke, Stuttgart

Hein JR, Koschinsky A (2014) Deep-Ocean Ferromanganese Crusts and Nodules. In: Treatise on Geochemistry: Second Edition, 2nd edn. Published by Elsevier Inc., pp 273–291

Hein JR, Spinardi F, Okamoto N et al (2015) Critical metals in manganese nodules from the Cook Islands EEZ, abundances and distributions. Ore Geol Rev 68:97–116. https://doi.org/10.1016/j.oregeorev.2014.12.011

Heye D, Marchig V (1977) Relationship between the growth rate of manganese nodules from the central Pacific and their chemical constitution. Mar Geol 23:M19–M25. https://doi.org/10.1016/0025-3227(77)90092-5

Hlawatsch S, Neumann T, Van Den Berg CMG et al (2002) Fast-growing, shallow-water ferromanganese nodules from the western Baltic Sea: origin and modes of trace element incorporation. Mar Geol 182:373–387. https://doi.org/10.1016/S0025-3227(01)00244-4

Josso P, Pelleter E, Pourret O et al (2017) A new discrimination scheme for oceanic ferromanganese deposits using high field strength and rare earth elements. Ore Geol Rev 87:3–15. https://doi.org/10.1016/j.oregeorev.2016.09.003

Karandashev VK, Turanov AN, Orlova TA et al (2008) Use of the inductively coupled plasma mass spectrometry for element analysis of environmental objects. Inorg Mater 44:1491–1500. https://doi.org/10.1134/S0020168508140045

Kasten S, Glasby GP, Schulz HD et al (1998) Rare earth elements in manganese nodules from the South Atlantic Ocean as indicators of oceanic bottom water flow. Mar Geol 146:33–52. https://doi.org/10.1016/S0025-3227(97)00128-X

Klevenz V, Sumoondur A, Ostertag-Henning C, Koschinsky A (2010) Concentrations and distributions of dissolved amino acids in fluids from Mid-Atlantic ridge hydrothermal vents. Geochem J 44:387–397. https://doi.org/10.2343/geochemj.1.0081

Kuhn T, Wegorzewski A, Ruhlemann C, Vink A (2017) Deep-sea mining. In: Sharma R (ed) Deep-sea mining: resource potential, technical and environmental considerations. Springer International Publishing, Cham, p 535

Kunzendorf H, Glasby GP, Stoffers P, Plüger WL (1993) The distribution of rare earth and minor elements in manganese nodules, micronodules and sediments along an east-west transect in the southern Pacific. Lithos 30:45–56. https://doi.org/10.1016/0024-4937(93)90005-W

Menendez A, James RH, Lichtschlag A et al (2019) Controls on the chemical composition of ferromanganese nodules in the Clarion-Clipperton fracture zone, eastern equatorial Pacific. Mar Geol 409:1–14. https://doi.org/10.1016/j.margeo.2018.12.004

Pattan JN, Parthiban G (2007) Do manganese nodules grow or dissolve after burial? Results from the Central Indian Ocean Basin. J Asian Earth Sci 30:696–705. https://doi.org/10.1016/j.jseaes.2007.03.003

Reykhard LY, Shulga NA (2019) Fe–Mn nodule morphotypes from the NE Clarion-Clipperton Fracture Zone, Pacific Ocean: comparison of mineralogy, geochemistry and genesis. Ore Geol Rev 110:102933. https://doi.org/10.1016/j.oregeorev.2019.102933

Skornyakova NS, Murdmaa IO (1992) Local variation in distribution and composition of ferromanganese nodules in the Clarion–Clipperton nodule province. Mar Geol 103:381–405

Skornyakova NS, Murdmaa IO (1995) Ferromanganese nodules and crusts in the Atlantic Ocean. Litol Polezn Iskop 30:339–361

Smetannikova OG (1988) Metodicheskie ukazaniya po rentgenovskomu issledovaniyu oksidov i gidroksidov marganca. Izdatelstvo LGU, Leningrad

Uspenskaya TY, Skornyakova NS (1991) Tekstury i struktury okeanskih zhelezo-margancevyh konkrecij i korok (in russian). Nauka, Moscow

Wegorzewski AV, Kuhn T (2014) The influence of suboxic diagenesis on the formation of manganese nodules in the Clarion Clipperton nodule belt of the Pacific Ocean. Mar Geol 357:123–138. https://doi.org/10.1016/j.margeo.2014.07.004

Wegorzewski AV, Kuhn T, Dohrmann R et al (2015) Mineralogical characterization of individual growth structures of Mn-nodules with different Ni + Cu content from the central Pacific Ocean. Am Mineral 100:2497–2508. https://doi.org/10.2138/am-2015-5122

Hiatuses and Core Correlations

Elena Ivanova, Dmitrii Borisov, and Ivar Murdmaa

Abstract The study of six sediments cores from the Ioffe Drift area documented the reduced thickness and/or absence of biostratigraphic zones and the occurrence of several hiatuses. The multi-proxy biostratigraphic, magnetic susceptibility (MS), color reflectance and X-ray fluorescence (XRF) data from five of the six cores reported numerous long- and short-term stratigraphic gaps over the last ~3–4 Ma. The correlation of sediment records from the drift suggests that the most pronounced series of hiatuses, associated with enhanced Lower Circumpolar Deep Water (LCDW) flow intensity, occurred from 2.51/2.59 to ~1.9 Ma (i.e., the onset of the modern-type deep-water circulation in the South Atlantic). This interval of specific high-amplitude peaks representing abrupt changes in volume MS and chemical composition variation may well serve as a regional stratigraphic benchmark in future studies of deep-sea contourites. A temporary intensification of the LCDW flow, probably due to its increased production in the Antarctic, led to deep erosion, ultimately resulting in long-term hiatuses and hence contributing to the enormously compressed Upper Pliocene–Middle Pleistocene section of the drift. The interval from 1.47/1.6 to 0.81 Ma, embracing the Mid-Pleistocene Transition, contains the longest stratigraphic gaps, up to ~1 Ma in some cores. Comparison of the studied sediment records to DSDP Site 516 reveals the reduced thickness of all the recovered biostratigraphic zones and evidence of more frequent hiatuses in the Ioffe Drift than on the neighboring Rio Grande Rise, suggesting more vigorous contour currents in the former.

E. Ivanova (✉) · D. Borisov · I. Murdmaa
Shirshov Institute of Oceanology, Russian Academy of Sciences, Moscow, Russia
e-mail: e_v_ivanova@ocean.ru

D. Borisov
e-mail: dborisov@ocean.ru

I. Murdmaa and E. Ivanova (eds.), *The Ioffe Drift*, Springer Geology,
https://doi.org/10.1007/978-3-030-82871-4_9

1 Definition and Identification of Hiatuses

As reported in our previous publications (Ivanova et al. 2016, 2020) and in Chaps. 6 and 7, our multi-proxy study revealed the widespread occurrence of hiatuses in all six sediment records available from the Ioffe Drift area (Figs. 1, 2, 3, 4, 5 and 6). In general, the hiatuses are not visible in the calcareous sediments of the Ioffe Drift during the visual description of the opened core sections; however, even if they are not straight or are masked by bioturbation (Fig. 7), in some cases the erosional contacts may be quite distinct.

Following Ivanova et al. (2016, 2020), the major hiatuses concerned here are ascertained or assumed geological phenomena expressed by the disappearance of

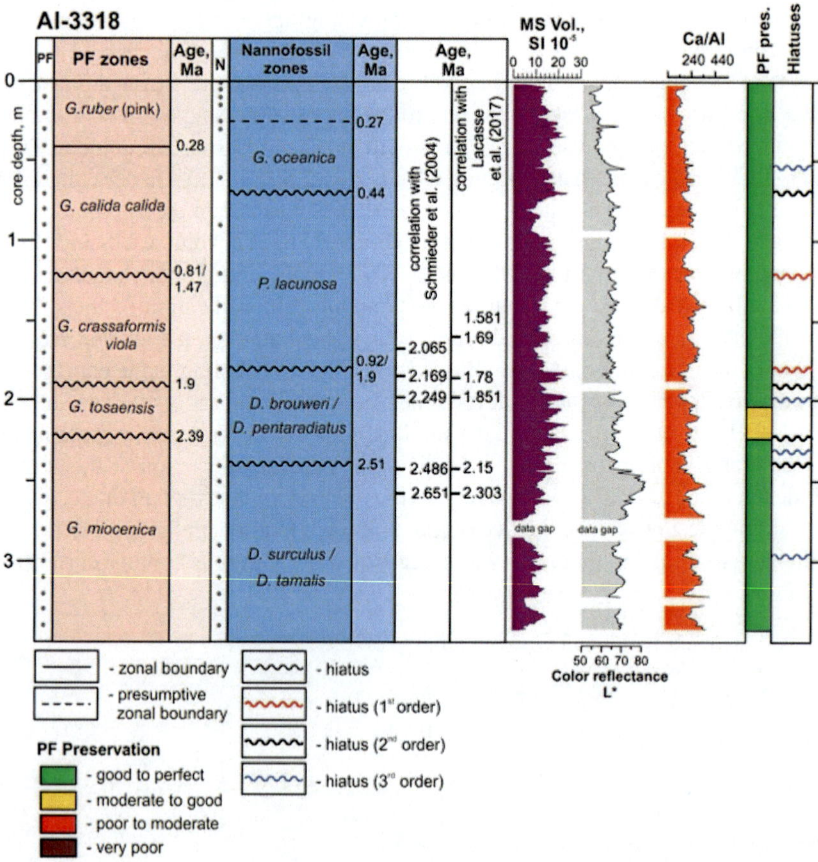

Fig. 1 Hiatuses of various orders in core AI-3318, inferred from biostratigraphy and down-core records of volume MS, color reflectance L* and Ca/Al. Planktic foraminiferal preservation (PF pres.) as well as the ages of MS peaks suggested by peak-to-peak correlation with Schmieder et al. (2004) and Lacasse et al. (2017) are also shown

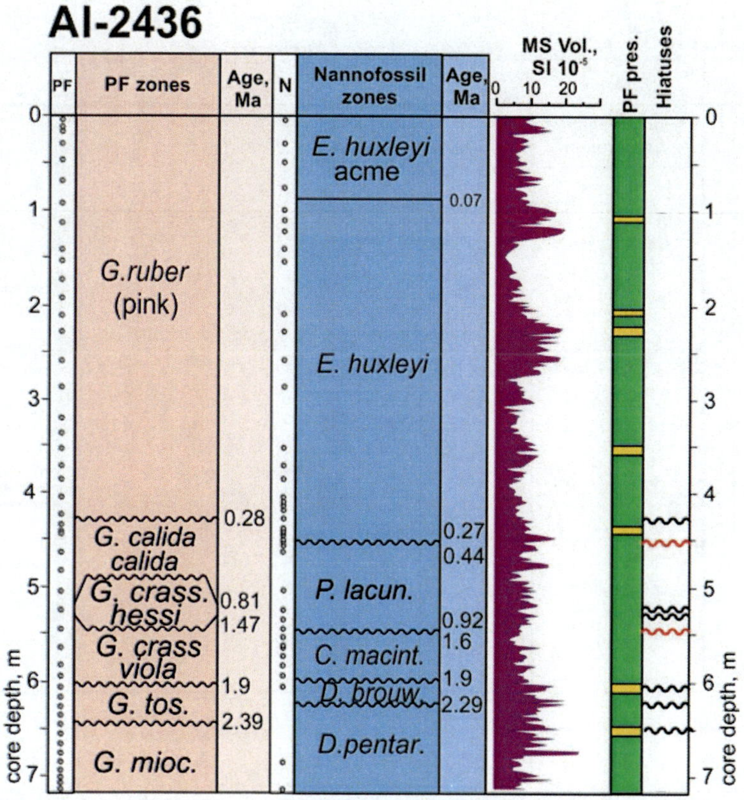

Fig. 2 Hiatuses of various orders in core AI-2436, inferred from biostratigraphy. Down-core record of volume MS and planktic foraminiferal preservation (PF pres.) are also shown. For legend, see Fig. 1

a stratigraphic interval from the sediment section (in this case, mainly contourite), irrespective of its physical origin, which might vary depending on sediment composition and paleoceanographic environments. The corresponding gap in biostratigraphic framework might be as long as several hundred thousand years (Figs. 1, 2 and 4). Such gaps correspond to hiatuses"of the first order". They can also be represented by seismic reflectors, like the boundary SU-3/SU-4, suggesting very deep erosion (see Chap. 5).

A hiatus "of the second order" is suggested by the unequivocally reduced thickness of any biostratigraphic zone relative to its thickness at other nearby locations, for instance on the Rio Grande Rise (see Chap. 7), or relative to the thickness of the other zones of similar duration in the same core section, In such cases, hiatuses are tentatively placed at the lower and upper zonal boundaries if their real position in the core section cannot be identified. The corresponding gap in the zone duration might be as long as several tens of thousands of years (Figs. 1, 2, 3, 4, 5 and 6). Sometimes, they are also characterized by changes in lithology as hiatuses of the"third order".

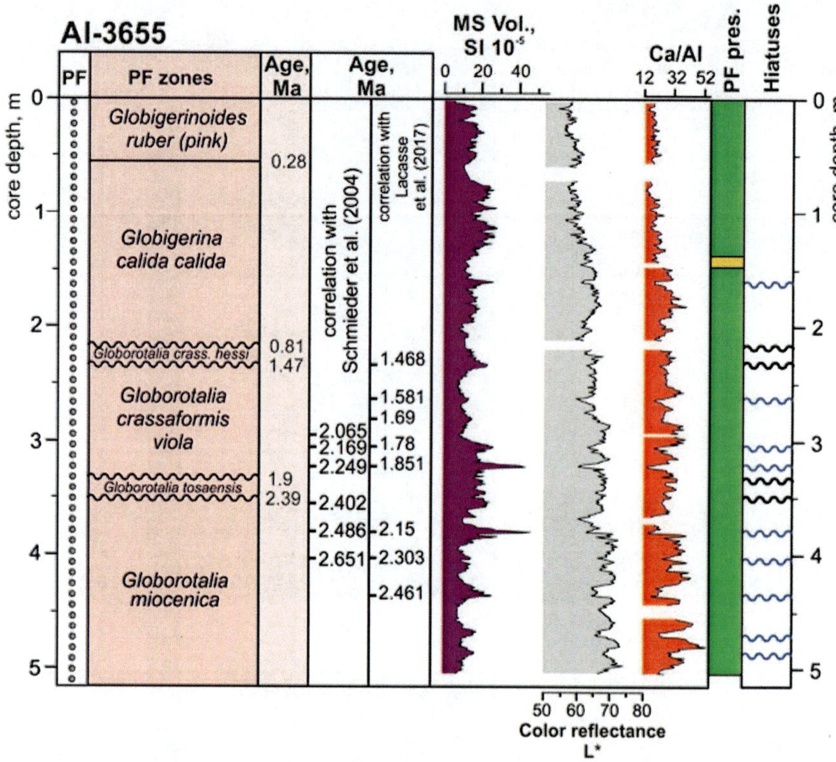

Fig. 3 Hiatuses of various orders in core AI-3365 inferred from biostratigraphy and down-core records of volume MS, color reflectance L* and Ca/Al. Planktic foraminiferal preservation (PF pres.) as well as the ages of MS peaks suggested by peak-to-peak correlation with Schmieder et al. (2004) and Lacasse et al. (2017) are also shown. For legend, see Fig. 1

As shown for our sediment cores AI-3318 and AI-3316 from the Ioffe Drift, several hiatuses are supported by high-amplitude changes in records of magnetic susceptibility (MS), X-ray fluorescence (XRF) and/or color reflectance (Ivanova et al. 2020). Similar variation patterns represented by an abrupt increase (or antiphase decrease) in certain parameters (such as MS, Si (norm.), Ca/Al ratio, or others) just after a hiatus followed by its gradual decrease (increase) are identified in all cores studied (Figs. 1, 2, 3, 4, 5, 6 and 8).

These changes discussed below indicate hiatuses"of the third order" in the sediment sections. The latter are probably instantaneous and the corresponding stratigraphic gaps, if any, are rather short (perhaps several hundred or a few thousand years). Most of the hiatuses of the third order correspond to distinctive irregular bioturbated contacts between dark brown layers relatively enriched with terrigenous material (high MS and Si(norm), low L* and Ca/Al ratio) and underlying light layers with relatively higher carbonate content (low MS and Si (norm), high L* and Ca/Al

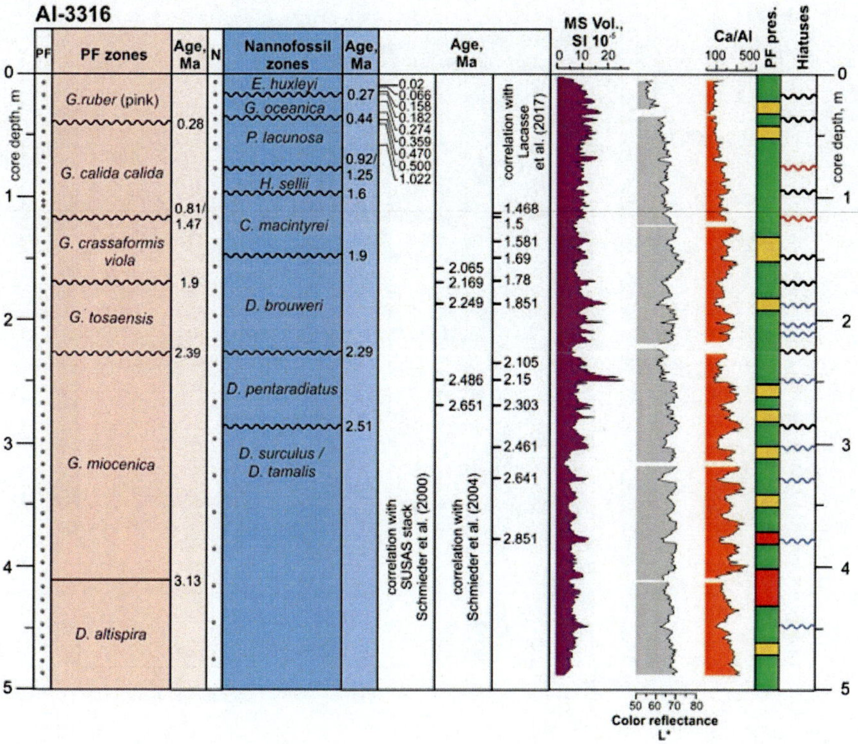

Fig. 4 Hiatuses of various orders in core AI-3316, inferred from biostratigraphy and down-core records of volume MS, color reflectance L* and Ca/Al. Planktic foraminiferal preservation (PF pres.) as well as the ages of MS peaks suggested by peak-to-peak correlation with Schmieder et al. (2000), Schmieder et al. (2004) and Lacasse et al. (2017) are also shown. For legend, see Fig. 1

ratio). Some of these contacts marking the third-order hiatuses are characterized by dark brown vertical burrows.

The sedimentological and hydrodynamic mechanisms of the hiatuses' formation need further investigation by the combined efforts of relevant specialists to achieve a better understanding. The authors believe that bottom-current activity acts as the major dynamic factor in the development of both erosional and non-depositional hiatuses. The"duration" or"timing" of a hiatus are simplified terms for the vanished stratigraphic time interval rather than the actual duration of the erosion (non-deposition) processes, which might be much shorter, perhaps even geologically instantaneous.

The geological correlation of long-term hiatuses between the cores studied by means of more reliable, yet low-resolution, zonal biostratigraphy and shorter-term data inferred from high-resolution MS, color reflectance and XRF records permits us to refine our stratigraphy of the uppermost acoustically stratified deposits in the central Ioffe Drift area (Figs. 1, 2, 3, 4, 5 and 6). In addition, the inter-correlation of several independent proxy records helps in the future elaboration of theoretical

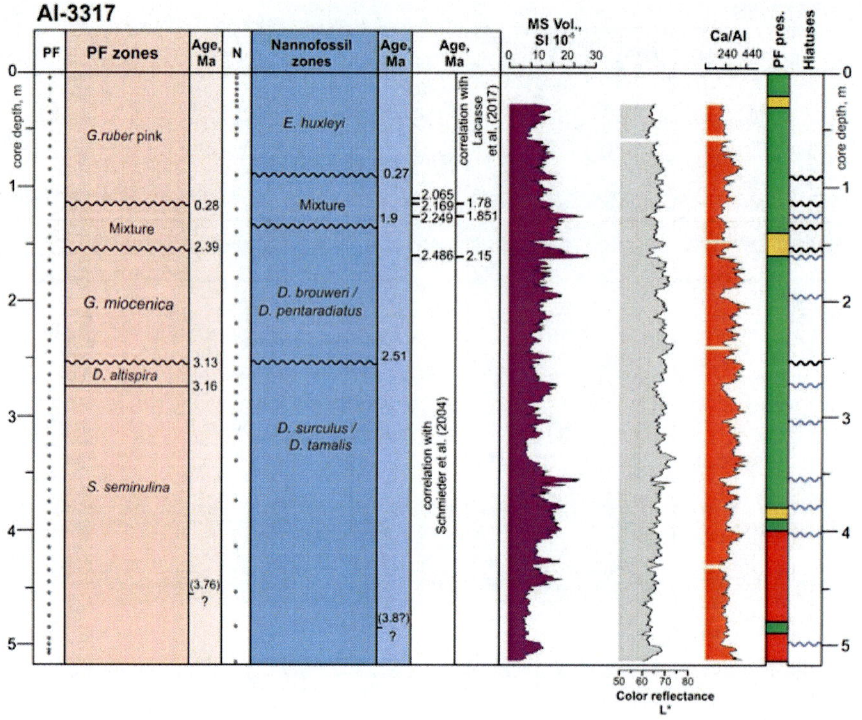

Fig. 5 Hiatuses of various orders in core AI-3317, inferred from biostratigraphy and down-core records of volume MS, color reflectance L* and Ca/Al. Planktic foraminiferal preservation (PF pres.) as well as the ages of MS peaks suggested by peak-to-peak correlation with Schmieder et al. (2004) and Lacasse et al. (2017) are also shown. For legend, see Fig. 1

approaches to explain the circulation patterns of the Antarctic Bottom Water (AABW) and Lower Circumpolar Deep Water (LCDW), both of which led to the discovery of the extensive formation of hiatuses so characteristic of the Ioffe Drift.

2 Multi-proxy Evidence and the Age of Hiatuses in the Upper Pliocene–Quaternary Sedimentary Cover of the Ioffe Drift

The series of hiatuses of the first-, second- and third-order were identified in five sediment cores retrieved from the Ioffe Drift area (Figs. 1, 3, 4, 5 and 6). In core AI-2436, collected in 2010, no material remained with which to obtain continuous XRF records and color reflectance measurements could not be made during the expedition. MS measurements were recently performed on this core at a resolution insufficient to identify hiatuses yet high enough to correlate to the other cores (Fig. 2; see also

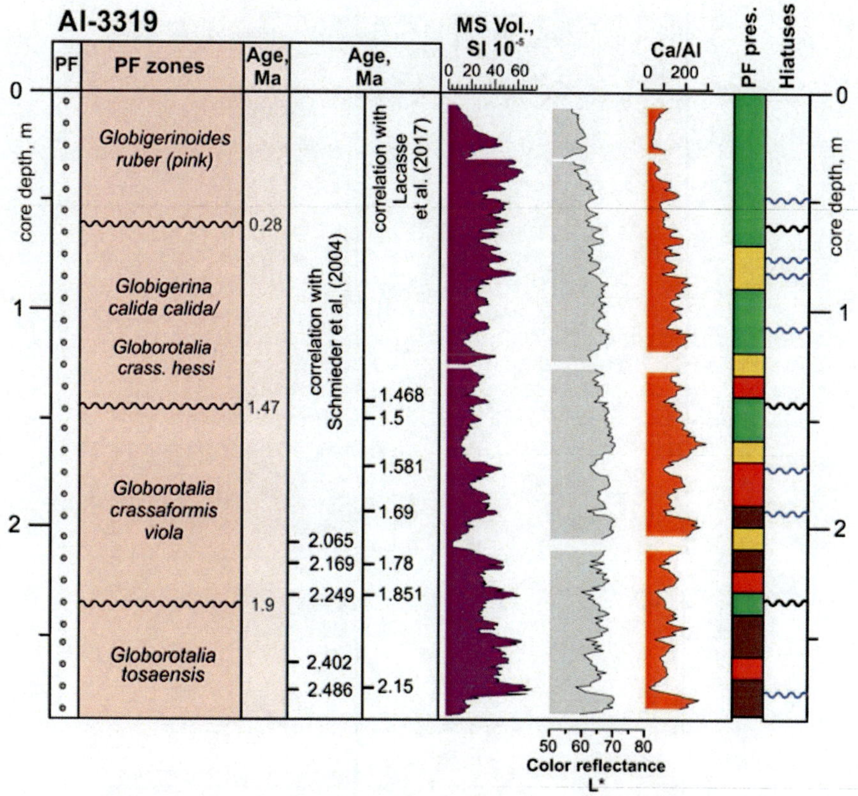

Fig. 6 Hiatuses of various orders in core AI-3318, inferred from biostratigraphy and down-core records of volume MS, color reflectance L* and Ca/Al. Planktic foraminiferal preservation (PF pres.) as well as the ages of MS peaks suggested by peak-to-peak correlation with Schmieder et al. (2004) and Lacasse et al. (2017) are also shown. For legend, see Fig. 1

Sect. "Geochemistry ($CaCO_3$, TOC, XRF)"); therefore, in this core the hiatuses are mostly reported through the biostratigraphy (Figs. 1 in Chap. 7 and 2; Ivanova et al. 2016).

In five sediment cores, the hiatuses were identified by three main parameters: (a) MS, reflecting the amount of terrigenous material in the drift deposits; (b) color reflectance (L*), characterizing the lightness of sediments (carbonate content, clay and nannofossil amount); and (c) Ca/Al ratio, demonstrating the relationship between the biogenic (calcareous) and lithogenic components of sediments. The other XRF data (i.e., Si, Al, Fe, Mn content) were generally used as supporting proxies. The abrupt simultaneous changes in these parameters allow more confident identification of hiatuses (Figs. 1, 3, 4, 5 and 6).

As mentioned above, the strongly asymmetric peaks, demonstrating an abrupt change followed by a gradual one, are considered a diagnostic criterion of a hiatus of the "third-order". These variation patterns were documented in all the cores studied

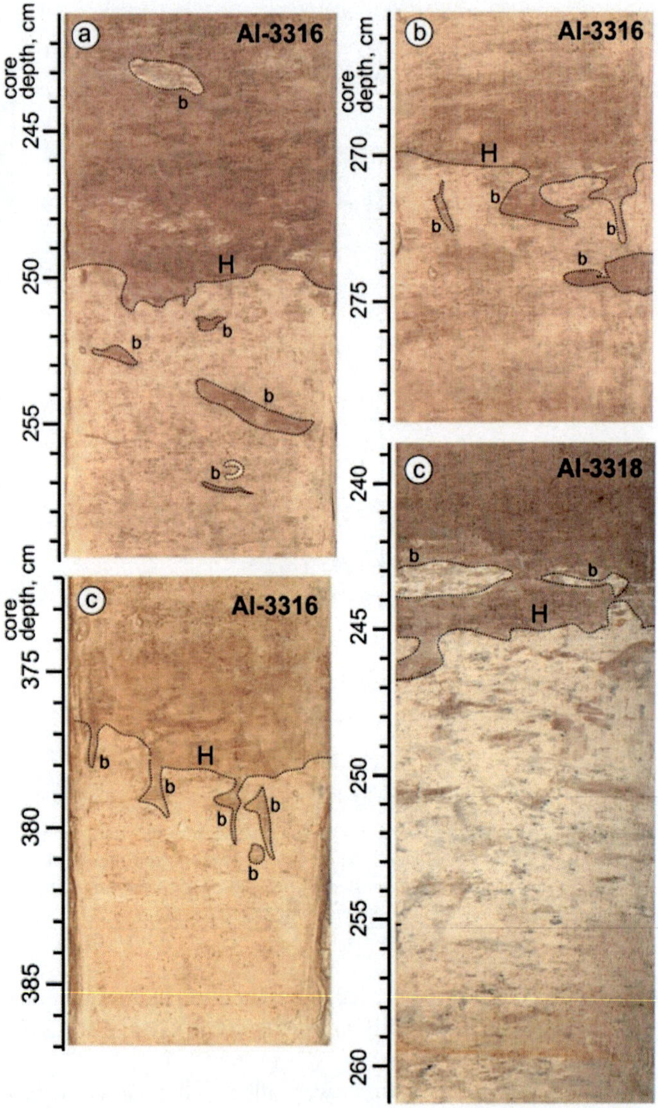

Fig. 7 Fragments of split-core photos; b–burrows, H–boundary, marking abrupt change in lithology that corresponds to a hiatus

(e.g., AI-3316: 235, 250, 330 and 380 cm; AI-3317: 92, 160 and 270 cm; AI-3655: 358, 380 and 436 cm; Figs. 8 and 9). They probably mark incomplete sedimentary cyclites, reflecting fluctuations of bottom-current velocity as described by Stow and Faugères (2008). The incompleteness of the cycles may have resulted from the drastic erosion that occurs when bottom flow reaches a maximum speed.

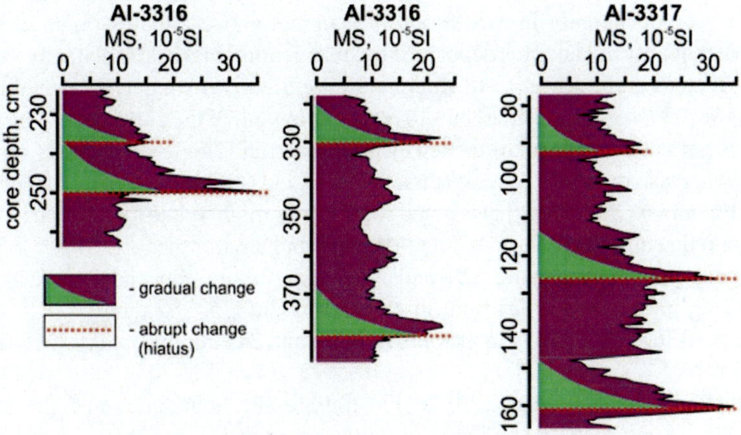

Fig. 8 Examples of abrupt volume MS changes from cores AI-3317 and AI-3316

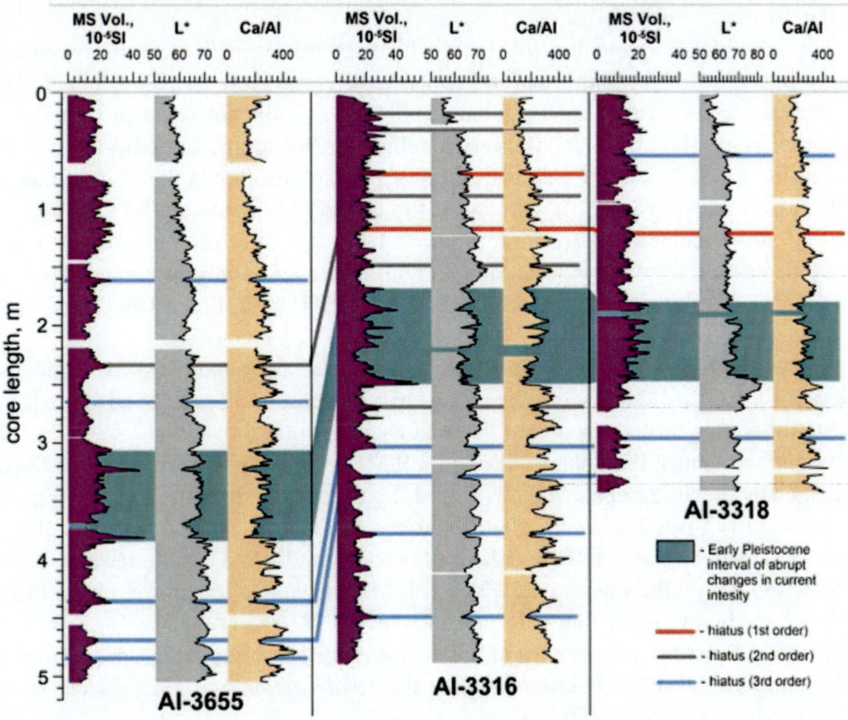

Fig. 9 Correlations of volume MS, color reflectance L* and Ca/Al records from cores AI-3316, AI-3655 and AI-3318

The abrupt increase in MS, Si and Al content above a hiatus (after an erosion event) probably means the deposition of terrigenous material, relatively enriched in Si, under conditions of a still-high bottom-current velocity. The gradual upward decrease in these parameters seems to reflect a slowing of the currents responsible for both the erosion and input of the terrigenous material. This assumption is supported by the increase in the Ca/Al ratio (Figs. 1, 3, 4, 5 and 6). The enrichment of sediments with Fe and especially Mn that is noted at MS peaks might relate to their concentration during oxic early diagenesis in a non-depositional environment.

Similarities between the MS and XRF records are especially evident in the following intervals: 170–247 cm in core AI-3316; 375–387 cm and 418–437 cm in core AI-3655; 269–289 cm in core AI-3317; and 243 cm in core AI-3318 (Figs. 1, 3, 4 and 5).

The frequent coincidence of the biostratigraphic boundaries and the MS and L* peaks in the records studied permits the more-or-less reliable identification of hiatuses. The most robust hiatuses (or series of hiatuses, as in the Lower Pleistocene sections) are nicely represented by abrupt changes in the MS and XRF records (Fig. 9).

In core AI-3318 from the drift summit, hiatuses at 120 cm and 180 cm correspond to an absence of foraminiferal and nannofossil zones, respectively (Fig. 1). The hiatus at 120 cm in foraminiferal zonation points to a significant gap in the Quaternary section from 1.47/1.6 to 0.81 Ma, very roughly corresponding to the Mid-Pleistocene Transition (MPT) and the intensification of bottom currents in the Vema Channel (Ivanova et al. 2016, 2020). Moreover, the underlying hiatus at 180 cm suggests a gap in the nannofossil zonation of about 1 Ma due to erosion and/or long-term non-deposition from 1.9 to 0.92 Ma (yet non-deposition over a million years seems unlikely). Similar hiatuses are reported from core AI-2436 (Figs. 1 in Chap. 7 and 2).

In core AI-3316, on the basis of the foraminiferal and nannofossil zonation, the levels 1.9 and 2.51/2.59 Ma were reported in both microfossil groups, with a reasonable difference in the down-core depth at each level (Ivanova et al. 2020; Fig. 4). Despite the hiatus, the younger level of 1.9 Ma seems to be robust and well documented by biostratigraphic and MS records in all cores apart from AI-3317 (Figs. 1, 3, 4 and 6). Several short-term erosional events may be inferred from the abrupt changes in the MS and XRF records in all cores (apart from AI-2436, see above, and AI-3319) within the interval 2.51/2.59–1.9 Ma, as well as the above, in the Pleistocene and below, in the Late Pliocene, in cores AI-3316 and AI-3317 (Figs. 1 3, 4, 5 and 6). In the latter two cores, the age of the lowermost hiatuses in the Upper Pliocene foraminiferal *Dentoglobigerina altispira* and *Sphaeroidinellopsis seminulina* zones can barely be identified (Figs. 4 and 5).

In core AI-3317, from the drift slope, the foraminiferal and nannofossil stratigraphy documents a ~40–50 cm interval with mixed faunal/floral complexes covering as much as 2.1 Ma and 1.6 Ma, respectively (Figs. 5 in Chap. 7 and 5). We argue that this interval also contains a series of hiatuses, which is nicely supported by the MS and XRF data (Fig. 5). As mentioned in Chap. 7, the discrepancy between the ages of the gaps inferred from foraminiferal and nannofossil zonations might be caused by the difference in the reworking of sand-size planktic foraminiferal tests by bottom currents compared to that of tiny and light coccoliths, as well as in the resistance of planktic foraminiferal and nannofossil taxa to selective dissolution. The latter was previously documented in DSDP/ODP Holes and sediment cores from nearby and remote locations (e.g., Berggren et al. 1983, 1995; Bylinskaya and Golovina 2004; Ivanova et al. 2016, 2020).

Previous studies have demonstrated the link between the location of ferromanganese nodules in sediment cores, hiatuses and bottom-current activity (e.g., Pautot and Melguen 1975; von Stackelberg 1979; Nishimura 1992). As the nodules are known to grow under low sedimentation rates (Kennett and Watkins 1975; Glasby 2006), those found at levels of stratigraphic hiatuses and abrupt changes in MS values (e.g., 226–228 and 247–248 cm in core AI-3318) might reflect a significant decrease in sedimentation rate and thus mark these hiatuses. However, the link between hiatuses and nodule occurrence was not confirmed in the other cores from the Ioffe Drift and thus needs further investigation.

Due to the strongly reduced thickness of biostratigraphic zones and the occurrence of multiple hiatuses and stratigraphic gaps, it is difficult to correlate our MS records to either the SUSAS stack (von Dobeneck & Schmieder 1999) or the available MS records from the subtropical South Atlantic (Schmieder 2004) and the shallow part of the Rio Grande Rise (Lacasse et al. 2017). The above-mentioned records cover approximately 1.5 Ma, 2.1 Ma and 3.5 Ma, respectively, thus they are of Pleistocene to Late Pliocene age at the base, similar to our cores. By contrast, the three records from von Dobeneck and Schmieder (1999), Schmieder (2004) and Lacasse et al. (2017) demonstrate high sedimentation rates, unlike in the Ioffe Drift area (see Chaps. 6 and 7).

Further, from the three cores studied, Lacasse et al. (2017) provides an age model based on $^{87}Sr/^{86}Sr$ dating of foraminiferal samples only for core MD11-L2P1. The authors document a rather variable number of MS peaks in the records from these three cores, indicating that some individual peaks cannot be identified and that their age cannot be ascertained, even by interpolation.

The upper six characteristic peaks (1.468, 1.5, 1.581, 1.69, 1.78, 1.851) from dated record MD11-L2P1 (Lacasse et al. 2017) can be rather well identified in the MS records of all cores from the Ioffe Drift apart from AI-3317. They generally fit the biostratigraphic frameworks to a reasonable extent (Figs. 1, 2, 3 and 5); however, below 1.9 Ma, a better correspondence of MS peaks to biostratigraphy is provided when the ages are obtained by MS correlate peak to peak with those from Schmieder (2004) for the lower to middle part of core AI-3318 (Fig. 1), as well as the lower parts of cores AI-3365 and AI-3316 (Figs. 3 and 4).

We consider three possible explanations of the discrepancy in the interpolated age of MS peaks and biostratigraphic age at the same levels in the lower part of sections, older than 1.9 Ma. One possibility is the stronger reworking of the Neogene species in the lower parts of our cores due to more vigorous bottom-water hydrodynamics on the Ioffe Drift summit prior to 1.9 Ma, notably in the Late Pliocene, than seems likely later on. Another reason might be the uncertainty over the identification of the MS peak ages in Lacasse et al. (2017), based on interpolation in a single record. The third possible explanation is the difference in sedimentation rates calculated for the Rio Grande Rise (Lacasse et al. 2017), almost ten times the very rough estimate for the biostratigraphic zones in the sections from the summit of the Ioffe Drift. As this difference is so high, we conclude that no reliable correlation of MS peaks can be achieved, at least not below 1.9 Ma; however, based on the available data, at present the other explanations cannot be ruled out.

The similar pattern of distinct high-amplitude peaks in MS, Ca/Al ratio and Ca/Ti ratio in the records, notably from cores AI-3318, AI-3655 and AI-3316 where they are best represented, allow the identification of the interval of strong erosion marked by multiple hiatuses in the earliest Pleistocene, from 2.51/2.59 to 1.9 Ma (Ivanova et al. 2020; Figs. 8 and 9). This conclusion is consistent with the calcareous biostratigraphy of the Ioffe Drift sediments described in Chap. 7. On the contrary, our earlier interpretation of the upper interval of the strong erosion, from 1.47/1.6 to 0.81 Ma (Ivanova et al. 2020), roughly embracing the MPT, is revised according to the new core-logging results (Fig. 9). The interval includes the hiatuses of the first- or second-order corresponding to the absence or reduced thickness of biostratigraphic zones in individual cores.

3 Paleoceanographic Implications of the Most Prominent Hiatuses

Our data demonstrate numerous hiatuses on the Ioffe Drift after ~3–3.5 Ma; that is, the onset of Pleistocene-type climatic and paleoceanographic fluctuations (Shackleton and Opdyke 1976; Vincent and Berger 1982; Berger and Wefer 1996) resulting in particular from the closure of the Central American (Panamanian) Seaway (e.g., Keigwin 1982; Schmittner et al. 2004; Ivanova et al. 2020). The closure was completed as the deep convection in the Northern Atlantic strengthened at 3.6 Ma (Haug and Tiedemann 1998). Some authors (e.g., Berger and Wefer 1996) suggest a third Neogene cooling step by 3 Ma and an associated pulsed increase in the activity of the bottom (AABW), deep (LCDW) and intermediate waters of Antarctic origin, as well as North Atlantic Deep Waters (NADW). According to Hodell and Venz (1992), the Late Pliocene interval from 3.2 to 3 Ma to ~2.5 Ma was rather unstable in the northern Antarctic and Subantarctic regions, with the formation of increased sea ice favoring deep convection. Recent studies have ascertained an overall strengthening of the Atlantic meridional overturning at 3–2.5 Ma (Karas et al. 2017) and changes in deep-water mass properties due to the Antarctic Sea ice advance (Hill et al. 2017).

The Pliocene/Pleistocene boundary at 2.51/2.59 Ma in our cores roughly corresponds to the onset of"modern" deep-water stratification in the South Atlantic (Turnau and Ledbetter 1989) after a major intensification of Northern hemisphere glaciation at 2.7 Ma ago (Haug and Tiedemann 1998; Lisiecki and Raymo 2005). We speculate that the reorganization of circulation and reintroduction of large volumes of AABW (and probably LCDW) into the Rio Grande Rise area (Turnau and Ledbetter

1989) might trigger stronger erosion of foraminiferal and nanno-foraminiferal ooze due to increased bottom-current activity and/or non-deposition in the study area, notably during the earliest Pleistocene, from 2.51/2.59 to 1.9 Ma. The earlier hiatuses on the Ioffe Drift cannot be dated precisely, yet one can assume that some correspond to increased paleospeed events in the Rio Grande area at 2.7, 2.85, 3.15–3.10 Ma and earlier, as estimated by Turnau and Ledbetter (1989).

A less-prominent interval from 1.47/1.6 to ~0.81 Ma, with several hiatuses and gaps in biostratigraphic zones (Fig. 9), probably reflects long-term erosion that ceased by the end of the MPT. The latter is defined at 0.95–0.8 Ma (e.g., Schmieder et al. 2000; de Garidel-Thoron et al. 2005; Kleiven et al. 2011). Earlier, Ledbetter & Cieselski (1986) suggested a reduction in the frequency and extent of deep hiatuses in the South Atlantic and southeast Indian Ocean by ~1 Ma. Strong erosion prior to 1–0.81 Ma seems to be associated with increased AABW (and probably LCDW) production and higher flow velocities of the deep western boundary currents suggested for glacials (Hall et al. 2001; McCave and Hall 2006; Ivanova et al. 2016).

4 Conclusions

The stratigraphic gaps created by hiatuses and documented in the foraminiferal and nannoplankton zonation of cores from the Ioffe Drift suggest that these hiatuses may have eliminated the sediments that had accumulated over tens and hundreds to hundreds of thousands of years; we recognize first-, second- and third-order hiatuses. The first are likely to have had a broader regional distribution and caused correspondingly long-term stratigraphic gaps (up to several hundred thousand years). Third-order hiatuses are nicely identified by abrupt changes in the MS and XRF records (sometimes also in color reflectance L) and can be correlated in several sediment cores.

The most prominent first- and second-order hiatuses are documented within the interval from 1.47/1.6 to ~0.81 Ma, probably reflecting long-term erosion that ceased by the end of the MPT.

Biostratigraphic data, MS, color reflectance L* and XRF records collectively permit the confident identification of numerous hiatuses in all six sediment cores studied from the Ioffe Drift area, notably within the interval from 2.5/2.59 to 1.9 Ma likely associated with the reorganization of deep-water circulation after the Pliocene/Pleistocene boundary. Some (probably short-lived) younger and older hiatuses are also tentatively identified.

The occurrence of hiatuses and the reduced thickness of all biostratigraphic zones in the Ioffe Drift sections compared to those at DSDP Site 516, on the nearby Rio Grande Rise, support our earlier suggestion about vigorous AABW contourite currents coming from the Vema Channel and strongly affecting the sedimentation (accumulation and erosion) in the drift area. It particularly concerns the intervals of pronounced change in global climate and ocean circulation during the Early Pleistocene.

References

Berger WH, Wefer G (1996) Expeditions into the past: paleoceanographic studies in the South Atlantic. In: Wefer G, Berger WH, Siedler G, Webb DJ (eds) The South Atlantic: present and past circulation. Springer, Berlin, pp 363–410

Berggren WA, Aubry MP, Hamilton N (1983) Neogene magnetobiostratigraphy of deep-sea drilling project, site 516: Rio Grande Rise, South Atlantic. In: Barker PF, Carlson RL, Johnson DA (eds) Initial reports of deep-sea drilling project 72. Government Printing Office, Washington, U.S, pp 675–713

Berggren WA, Hilgen FJ, Langereis CG et al (1995) Late Neogene chronology: new perspectives in high-resolution stratigraphy. Geol Soc Am Bull 107:1272–1287. https://doi.org/10.1130/0016-7606(1995)107%3c1272:LNCNPI%3e2.3.CO;2

Bylinskaya ME, Golovina LA (2004) Correlation of the Pliocene-quaternary foraminiferal and nannofossil zonations in the North Atlantic. Stratigr Geol Correl 12(3):309–319 (in Russian with English trans)

de Garidel-Thoron T, Rosenthal Y, Bassinot FC, Beaufort L (2005) Stable sea surface temperatures in the western Pacific warm pool over the past 1.75 million years. Nature 433:294–298. https://doi.org/10.1038/nature03189

Glasby GP (2006) Manganese: predominant role of nodules and crusts. In: Shulz HD, Zabel M (eds) Marine geochemistry. Springer, pp 371–427 https://doi.org/10.1007/3-540-32144-6_11

Hall IR, McCave IN, Shackleton NJ et al (2001) Intensified deep Pacific inflow and ventilation during Pleistocene glacial times. Nature 412:809–812. https://doi.org/10.1038/35090552

Haug GH, Tiedemann R (1998) Stable carbon and oxygen isotope ratios of *Cibicidoides wuellerstorfi*, and $CaCO_3$ and sand content of ODP Hole 165–999A. Pangaea. https://doi.org/10.1594/PANGAEA.789866

Hill DJ, Bolton KP, Haywood AM (2017) Modelled ocean changes at the Plio-Pleistocene transition driven by Antarctic ice advance. Nat Commun 8:14376. https://doi.org/10.1038/ncomms14376

Hodell DA, Venz K (1992) Toward a high-resolution stable isotopic record of the Southern Ocean during the Pliocene–Pleistocene (4.8–0.8MA). In: Kennett JP, Warnke DA (eds) The antarctic paleoenvironment: a perspective on global change part 1, vol 56 (Antarctic Research Series). American Geophysical Union, Washington DC, pp 265–310

Ivanova E, Murdmaa I, Borisov DG et al (2016) Late Pliocene-Pleistocene stratigraphy and history of formation of the Ioffe calcareous contourite drift, Western South Atlantic. Mar Geol 372:17–30. https://doi.org/10.1016/j.margeo.2015.12.002

Ivanova E, Borisov D, Dmitrenko O, Murdmaa I (2020) Hiatuses in the late Pliocene-Pleistocene stratigraphy of the Ioffe calcareous contourite drift, western South Atlantic. Mar Pet Geol 111:624–637. https://doi.org/10.1016/j.marpetgeo.2019.08.031

Karas C, Nürnberg D, Bahr A et al (2017) Pliocene oceanic seaways and global climate. Nat Sci Rep 7:39842. https://doi.org/10.1038/srep39842

Keigwin LD (1982) Isotope paleoceanography of the Caribbean and east Pacific: role of panama uplift in late Neogene time. Science 217:350–353

Kennett JP, Watkins ND (1975) Deep-sea erosion and manganese nodule development in Southeast Indian Ocean. Science 188:1011–1101

Kleiven HF, Hall IR, McCave IN et al (2011) Coupled deep-water flow and climate variability in the middle Pleistocene North Atlantic. Geology 39(4):343–346. https://doi.org/10.1130/G31651.1

Lacasse CM, Santos RV, Dantas EL et al (2017) $^{87}Sr/^{86}Sr$ dating and preliminary interpretation of magnetic susceptibility logs of giant piston cores from the Rio Grande Rise in the South Atlantic. J S Am Earth Sci 80:244–254. https://doi.org/10.1016/j.jsames.2017.09.034

Ledbetter MT, Ciesielski PF (1986) Post-miocene disconformities and paleoceanography in the Atlantic sector of the Southern Ocean. Palaeogeogr Palaeoclim Palaeoecol 52(3–4):185–194, 197–214

Lisiecki LE, Raymo ME (2005) A Pliocene–Pleistocene stack of 57 globally distributed benthic $\delta^{18}O$ records. Paleoceanography 20:PA1003. https://doi.org/10.1029/2004PA001071

McCave IN, Hall IR (2006) Size sorting in marine muds: processes, pitfalls, and prospects for paleoflow-speed proxies. Geochem Geophys Geosyst 7:Q10N05. https://doi.org/10.1029/2006GC001284

Nishimura A (1992) Sedimentation and hiatuses in the Central Pacific Basin: their relationship to manganese nodule formation. In: Keating DH, DolLon DR (eds) Geology and offshore mineral resources of the central Pacific Basin, Circum-Pacific Council for energy and mineral resources earth science series 14. Springer, New York

Pautot M, Melguen M (1975) Deep bottom currents, sedimentary hiatuses and polymetallic nodules. In: I.D.O.E Workshop, 1975–09, pp 227–234

Schmieder F, von Dobeneck T, Bleil U (2000) The Mid-Pleistocene climate transition as documented in the deep South Atlantic Ocean: initiation, interim state and terminal event. Earth Planet Sci Lett 179:539–549. https://doi.org/10.1016/S0012-821X(00)00143-6

Schmieder F (2004) Magnetic signals in Plio–Pleistocene sediments of the South Atlantic: chronostratigraphic usability and paleoceanographic implications. In: Wefer G, Mulitza S, Ratmeyer V (eds) The South Atlantic in the late quaternary: reconstruction of material budgets and current systems. Springer, pp 263–279

Schmittner A, Sarnthein M, Kinkel H et al (2004) Global impact of the Panamanian Seaway closure. Eos 85:526. https://doi.org/10.1029/2004EO490010

Shackleton NJ, Opdyke ND (1976) Oxygen-isotope and paleomagnetic stratigraphy of Pacific core V28–239, Late Pliocene to latest Pleistocene. Geol Soc Am Mem 145:449–464

Stow DAV, Faugères J-C (2008) chapter 13. Contourite facies and the facies model. In: Rebesco M, Camerlenghi A (eds) Contourites. Developments in sedimentology, 60. Elsevier, Amsterdam, pp 223–256. https://doi.org/10.1016/S0070-4571(08)10013-9

Turnau R, Ledbetter MT (1989) Deep circulation changes in the South Atlantic Ocean: response to initiation of northern hemisphere glaciation. Paleoceanography 4:565–583

Vincent E, Berger WH (1982) Planktonic foraminifera and their use in paleoceanography. In: Emiliani C (ed) The Sea, vol 7. Wiley-Interscience, New York

von Dobeneck T, Schmieder F (1999) Using rock magnetic proxy records for orbital tuning and extended time series analyses into the super- and sub-Milankovitch bands. In: Fischer G, Wefer G (eds) Use of proxies in paleoceanography: examples for the South Atlantic. Springer, Berlin Heidelberg, pp 601–633

von Stackelberg U (1979) Sedimentation, hiatuses, and development of manganese nodules: VALDIVIA Site Va-13/2, Northern Central Pacific. In Bischoff JL, Piper DZ (eds) Marine geology and oceanography of the pacific manganese nodule province. Springer US, Boston, MA, pp 559–586. https://doi.org/10.1007/978-1-4684-3518-4_16

History of the Ioffe Drift

Ivar Murdmaa, Elena Ivanova, and Dmitrii Borisov

Abstract The comprehensive analysis of a large dataset of seismic, lithological, geochemical and micropaleontological data has shed light on the Ioffe Drift's origin and evolution. Although the drift's prehistory still has blank spots, it is suggested that oceanographic processes may have affected the formation of this depositional body since its earliest stage. The drift's formation probably began over the volcanic ridge in the Florianopolis Fracture Zone just after the ridge's formation approximately 95–80 Ma ago. Variations in bottom-current intensity in response to regional and global paleoceanographic and paleoclimatic changes are most clearly imprinted in the seismic structure of the upper 80 m of drift deposits and in sediment records recovered by cores (down to 7 m below sea floor (mbsf)). The drift's contourite formation underwent alternating phases of erosion caused by the intensification of bottom currents, probably followed by prevailing pelagic settling upon weakening of the bottom currents. The most thoroughly studied history of the Ioffe contourite drift, the Late Pliocene to Recent, was interrupted by multiple erosional hiatuses resulting from the activity of the Lower Circumpolar Deep Water (LCDW) bottom current during the intervals from 2.51/2.59 to 1.9 Ma and from 1.47/1.6 to 0.81 Ma. The numerous hiatuses that detached layers of calcareous contourites are an essential characteristic of the Ioffe Drift deposits, as well as layered sedimentary structures expressing effects of different bottom-currents velocities on the contourite deposition.

I. Murdmaa · E. Ivanova (✉) · D. Borisov
Shirshov Institute of Oceanology, Russian Academy of Sciences, Moscow, Russia
e-mail: e_v_ivanova@ocean.ru

I. Murdmaa
e-mail: murdmaa@mail.ru

D. Borisov
e-mail: dborisov@ocean.ru

I. Murdmaa and E. Ivanova (eds.), *The Ioffe Drift*, Springer Geology,
https://doi.org/10.1007/978-3-030-82871-4_10

1 Introduction

A realistic geological and paleoceanographic history of the Ioffe Drift can be reconstructed only for the uppermost section, recovered by six gravity cores (3.5–7 mbsf), spanning about 3–4 Ma BP (Before Present), on which multidisciplinary lithological, micropaleontological, biostratigraphic and multi-sensor core-logging studies were carried out. Below this recovered interval, only high-resolution seismic records (to a maximum of 80 mbsf) or a single-channel low-resolution seismic profile (down to 225 mbsf on the ridge summit) are available to interpret the geological history. Therefore, our consideration of the major lower part of the drift is much more speculative. We refer to it as a prehistory mainly based on hypotheses.

2 Early Prehistory of the Ioffe Drift

We know about the earliest stages of the Ioffe Drift's growth over the Florianopolis Fracture Zone Ridge only from a single-channel (air-gun) low-resolution seismic profile across the drift from northeast to southwest (see Fig. 2 in Chap. 5, GeoMapApp: www.geomapapp.org). This provides a rather rough two-dimensional image of the basement surface and major acoustic reflectors within the overlying sedimentary body. Its growth apparently commenced after the cessation of the intraplate (transform fault) volcanism that built the ridge, possibly with a stratigraphic hiatus at the base of the sedimentary body, but no earlier. Therefore, the reconstructed plate tectonic age of the volcanic Florianopolis Fracture Zone Ridge in the study area gives a maximum age for the sedimentary drift of about 95–80 Ma, according to Müller et al. (1997).

The structure, composition and age of the lower drift section are documented only by the low-resolution seismic profile, thus cannot be adequately interpreted; however, the shape of the basement reflector demonstrates infilling of a local basement depression by deposits with northward sloping reflectors (see Fig. 2 in Chap. 5). The configuration of these reflectors indicates that sediments continued to accumulate after the depression was infilled, showing the growth of a new sedimentary body over the basement's surface that morphologically resembles a mounded contourite feature (Fig. 1a). So far as the inclusion of any new terrigenous material source seems unlikely in this assumed geomorphologically isolated pelagic realm, we hypothesize that the only sedimentary material source was biogenic, probably calcareous planktonic, like that which has continuously accumulated on its neighbor, the Rio Grande Rise at least since the Santonian, according to DSDP results at Site 516 (Barker et al. 1983a).

If so, we can assume that fragmental biogenic calcareous deposits started to accumulate on the Florianopolis Ridge soon after or even simultaneously with the last basaltic eruptions. The basement surface morphology revealed by the seismic profile

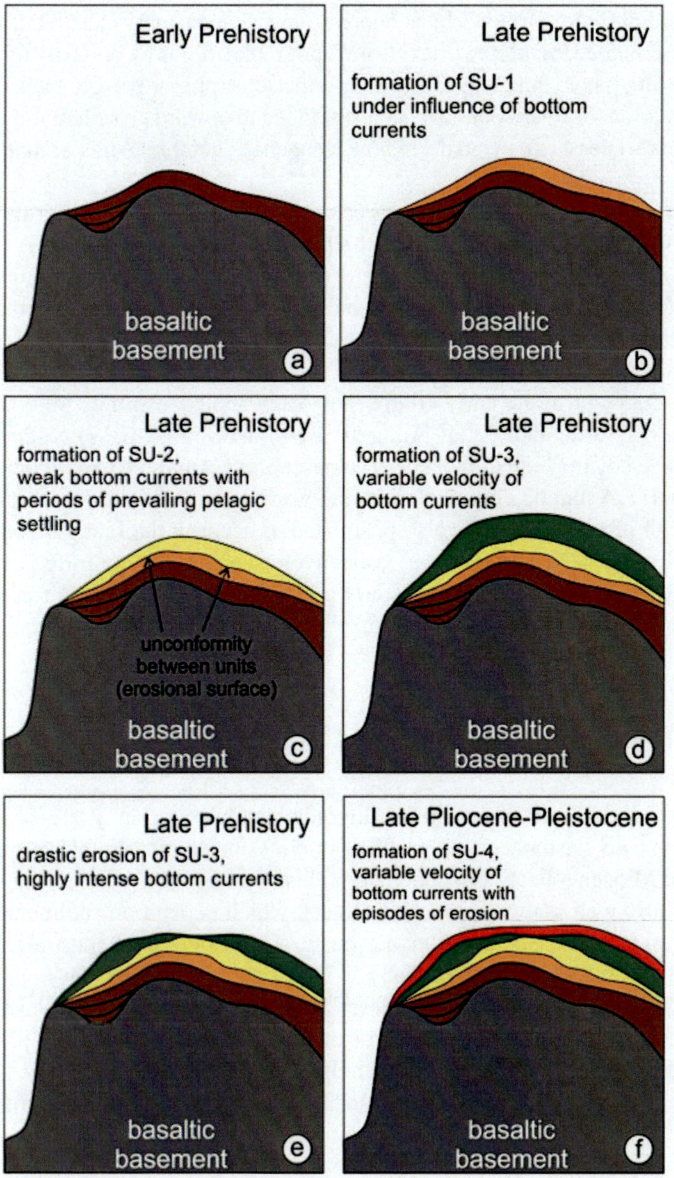

Fig. 1 a–f Sketch representing the stages of the Ioffe Drift formation (see Fig. 2 in Chap. 5) shows local depressions where primary calcareous ooze probably accumulated, being reworked from topographic highs by bottom-currents hence representing patchy infilling contourites rather than proper pelagic facies of vertical sedimentation. Further contourite deposition led to upward growth of the future Ioffe Drift and its lateral extension during the Neogene at depths shallower than the calcite compensation depth (CCD).

From a paleoceanographic perspective, deep-to-bottom water formation in the Southern Ocean has occurred at least since the late Paleogene; that is, it possibly turned on even prior to the opening of the Drake Passage, according to several studies (e.g., Scher and Martin 2008, and references therein). Some authors believe the modern-like Atlantic overturning circulation started with a proto-North Atlantic Deep Water (NADW) formation as early as the late Eocene, at about 35 Ma (Via and Thomas 2006) or in the early Oligocene (Allen and Armstrong 2008). Pfuhl et al. (2004) argue for an increase in Antarctic Bottom Water (AABW) production in the Weddell Sea by the Late Oligocene. The onset of the Antarctic Circumpolar Current, isolating the Antarctic continent from the warm ocean and thereby contributing to the growth of its ice sheet, is also postulated to occur in the Late Oligocene (Pfuhl and McCave 2005). These studies collectively suggest that the Ioffe Drift area has probably been influenced since its early prehistory by AABW and/or deep water of southern origin. This influence presumably caused the onset of contourite deposition.

3 Late Prehistory of the Ioffe Drift

Although direct data on sediment lithology and stratigraphy are absent, high-resolution SES seismoacoustic profiling provides much more information about the probable Miocene–Early Pliocene late prehistory of the Ioffe Drift (Fig. 1b). As is theoretically well known, screening effects, which depend on sediments' acoustic properties, restrict the penetration of acoustic signals into sediment strata. Therefore, the relative thickness of strata that are acoustically visualized by high-resolution seismic records may indicate the presence of such features as sand layers intercalated into a section. Thick intervals of the acoustic recovery recorded by SES are thus interpreted as an absence of screening effects in the Ioffe Drift, since maximum acoustic signal penetration reaches up to 80 m on the drift top. Thinning of the acoustically recovered intervals indicates the influence of acoustic signal screening (see Figs. 3, 4 and 5 in Chap. 5).

As shown in Chap. 5, four seismic units can be distinguished in the upper sediment cover of the drift. These reflect three sedimentation modes as considered below. Based on the acoustic structure and thickness of the inferred seismic units, we suggest that acoustically transparent SU-2 was deposited under conditions of a weak but stable current, probably with periods of prevailing pelagic settling (Fig. 1b–e).

Acoustically stratified units SU-1, SU-3 and SU-4 reflect more intense and variable bottom-water dynamics with episodes of drastic erosion. The deposition of SU-3 was affected by the most intense bottom currents, resulting in the deepest erosion

(Fig. 1e). The thickness distribution of unit SU-4 demonstrates the location of the well-established depocenter on the drift top, embraced by a local gyre (as inferred from the numerical modeling data (Ivanova et al. 2020)). The higher sedimentation rates in the center of the gyre probably resulted in the increased thickness of the unit on the drift's summit. For details on each unit's structure and thickness, see Chap. 5.

As mentioned above, we cannot date the late prehistory of the Ioffe Drift. The sediment cover represented by the units SU-1 to the lower part of SU-4 was most likely deposited in the Miocene–Early Pliocene. The reflector truncation, with a significant decrease in unit thickness (see e.g., Fig. 3a–c in Chap. 5) and irregular shape of boundaries between units (see Fig. 5a, b in Chap. 5), suggests that the unit boundaries represent a result of drastic erosion caused by bottom-current activity. This is the only process able to produce such erosional surfaces on the scale of the entire drift. The dramatic increase in bottom-current velocities probably occurred in response to regional or global changes in ocean circulation and climate. This assumption implies that evidence of these erosion events can be found in the sediment records from areas adjacent to the drift (e.g., the Rio Grande Rise). However, due to the insufficient data, we can only tentatively estimate the age of the boundaries between the seismic units.

Formation of the upper Ioffe Drift and corresponding seismic units would be affected by major paleoceanographic events such as the increase in Southern Component Water production (Hall et al. 2003) associated with the Middle Miocene cooling at around 14 Ma and major growth of the East Antarctic Ice Sheet from 15.5 to 13.5 Ma (Shackleton and Kennett 1975; Berger and Wefer 1996). Another important paleoceanographic event seems to be the Late Miocene (10.6 and 7.3 Ma) change in deep-water stratification in the Atlantic suggested by εNd data from a depth transect on the Walvis Ridge (Thomas and Via 2007). This event is also supported by the differences in $\delta^{13}C$ patterns which evidence the age difference between the abyssal waters of the South Atlantic and South Pacific as a result of turning on the NADW formation in the North Atlantic (Berger and Wefer 1996).

In DSDP Sites 516 and 518 from the Rio Grande Rise, a prominent erosional hiatus is reported in the Late Miocene (Barker et al. 1983a, b). The angular unconformity between units SU-4 and SU-3 (see Figs. 3, 4 and 5 in Chap. 5) reflects the most intense and the deepest erosion of the entire geological time interval covered by the high-resolution seismoacoustic records. It might be coeval to the gap of ~1–1.5 Ma documented in the sedimentary record of Site 518 near the Miocene/Pliocene boundary, at 5.5–4 Ma, presumably related to an intensification of abyssal circulation in the Vema Channel (Barker et al. 1981). By the end of Miocene, the interoceanic deep-water exchange via the gradually closing Central American Seaway was significantly reduced, and the modern pattern of deep-water circulation evolved in the Atlantic (Nisancioglu et al. 2003; Ivanova 2009). The NADW flow to the South Atlantic intensified, supported by the inflow of Mediterranean waters and the merid ional heat transfer to the North Atlantic (Haug and Tiedemann 1998; Haq et al. 1987; Ivanova 2009).

The delivery of heat to the Southern Ocean and its extraction from the South Atlantic also increased (Berger and Wefer 1996). We speculate that these paleoceanographic changes are reflected in the boundary between units SU-3 and SU-4, representing evidence of the deepest erosion in the Ioffe Drift area. Although the lengths of the sediment cores from the Ioffe Drift recovering the last ~3–4 Ma (see Chap. 7) comprise only a half of the uppermost unit SU-4 thickness, we speculate that the boundary between SU-3 and SU-4 can be tentatively assigned to the Late Miocene–Early Pliocene transition bearing in mind the aforementioned hiatus on the neighboring Rio Grande Rise. The hiatus is likely associated with the Late Miocene cooling and corresponding changes in ocean circulation (Hodell and Kennett 1985; Krasheninnikov and Basov 1986; Ivanova et al. 1989).

4 Late Pliocene to Recent History of the Ioffe Drift

Multidisciplinary studies of the sediment cores retrieved from the Ioffe Drift summit (AI-2436, AI-3318, AI-3655) and upper slope (AI-3316) characterize its uppermost sediment cover as a high-calcareous nannofossil-foraminiferal contourite (Fig. 1f; see Chap. 6). This conclusion concerns the upper part of 4–21 m (about 10 m on average) of the seismic unit SU-4 with thin parallel stratification. According to the biostratigraphy, five core sections penetrated the entire Quaternary and entered into the Upper Pliocene deposits below (see Chap. 7). Uniform seismic pattern of the whole SU-4 permits us to suggest that lithology of its lower part not reached by the cores is generally similar to that recovered by the cores and described in Chap. 6. The recovered Upper Pliocene deposits including the stratigraphically oldest penetration (up to 4 Ma) in core 3317 are characterized by a relatively high proportion of nannofossils and very low foraminiferal sand content (see Figs. 1, 2, 3, 4, 6 in Chap. 6 and Figs. 1, 2, 3, 4 and 5 in Chap. 7). This is likely related to a weak bottom-water dynamics (see in Sect. "Paleoceanographic Context of the History of the Ioffe Drift") rather than to more abundant nannofossils supply from their primary production.

Along with seismic facies, the contourite origin of the drift was ascertained by the sedimentary structures, grain-size distribution, numerous erosional hiatuses inferred from the zonal biostratigraphy and the abrupt changes in down-core volume magnetic susceptibility (MS), color reflection and X-ray fluorescence (XRF) scanning values (Ivanova et al. 2020, see also Chaps. 5–7 and 9). A good correlation of the sortable silt (*SS*) content in the total <63 μm size fraction and *SS* mean sizes inferred in four of six cores studied from the drift area (Fig. 10 in Chap. 6) confirms this suggestion. The intervals with frequent hiatuses cover a significant part of the roughly 3–4 Ma-long latest history of the Ioffe Drift or may even prevail throughout depositional time intervals (Fig. 2; Ivanova et al. 2020).

The impact of gravity-driven processes on the Ioffe Drift formation is negligible mostly due to its location and morphology. Bioturbation, thought to be a sedimentary structure typical of contourites (Wetzel et al. 2008; Stow et al. 2019), indicates slow and continuous deposition in contrast to geologically instantaneous turbidites and

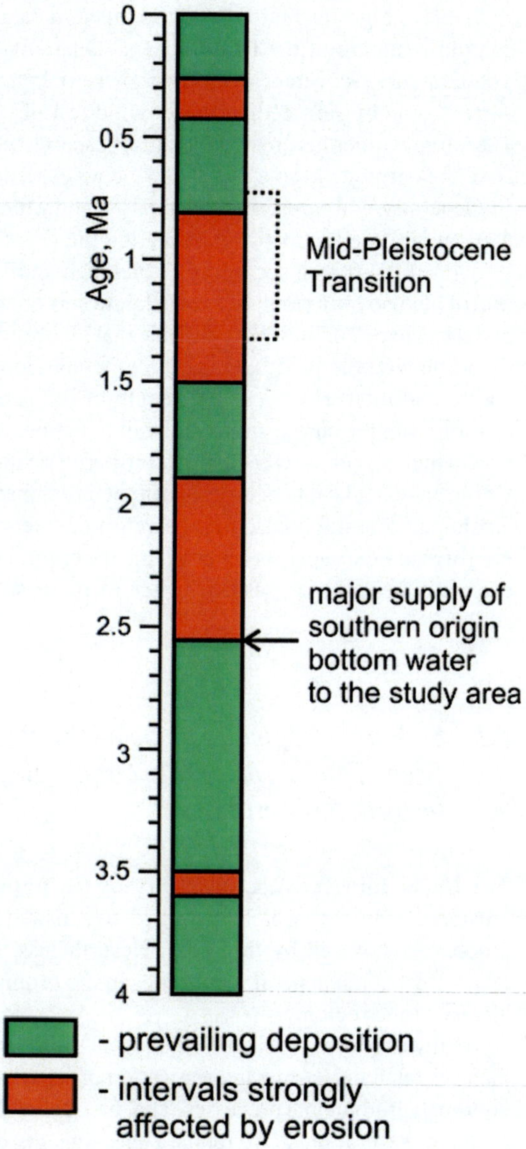

Fig. 2 Intervals of prevailing sediment accumulation and widespread erosion by vigorous bottom currents during the Late Pliocene–Recent history of the Ioffe Drift (from Ivanova et al. (2016) with modifications and additions). The input of the southern-origin bottom water is shown according to Turnau and Ledbetter (1989)

other types of gravitites. It is rather well expressed in all the cores studied. Uneven boundaries of darker clayey interlayers were probably disturbed by macro-benthos activity, and a slightly increased content of organic matter settled together with clay suggests a possible acceleration. However, bioturbation is also characteristic of vertical pelagic (hemipelagic) sedimentation (Bard et al. 1987; Trauth et al. 1997), thus it cannot be used as a reliable criterion specifically of contourites.

Overall, high resolution seismoacoustic data and lithology of cores support the contourite nature of the Late Pliocene–Quaternary development of the Ioffe Drift. The contourite-covered areas are shown to experience high and variable bottom-current velocities exceeding those in the rest of the ocean (e.g., Thran et al. 2018). Therefore, numerous erosional hiatuses ascertained in all core records from the Ioffe Drift area provide even a more convincing evidence of bottom-current activities.

Generally well-preserved planktic foraminiferal (PF) tests, notably in the upper parts of all core sections and indicating the cores' location above the foraminiferal lysocline (at 4,050 m according to Melguen and Thiede (1974)) exclude the formation of hiatuses as a result of total calcium carbonate dissolution (see Figs. 1, 2, 3, 4, 5 and 6 in Chap. 7). Thus, we conclude that all observed hiatuses resulted from erosion by oceanographic processes including along-slope contour currents, tidal currents, eddies and internal waves. The length of the geological time interval removed by erosion from the chronostratigraphy of a section does not equate to the erosion duration, which might have been much shorter, perhaps even geologically instantaneous. Unfortunately, the rate of erosion cannot be estimated; any hiatus corresponding to an erosional event might also mask a vanished interval of sediment deposition. Only if microfossil index species are available the chronostratigraphic age of the last erosion event can be estimated using the age of the basal sediment deposited at the hiatus surface.

4.1 Sedimentological Effects of the Bottom Circulation: From Total Erosion Through Selective Deposition to Bulk Accumulation

The Upper Pliocene–Recent history of the Ioffe Drift represents a succession of erosional hiatuses, which disturbed a considerable portion of the stratigraphic sequence recovered by the cores. Nevertheless, depositional residues allow us to reconstruct some general regularities in the sedimentological and thus hydrological history of the region.

As it was shown in Chap. 9, the Ioffe Drift is characterized by common erosional hiatuses related to strong intensification of (contour) bottom currents and other near-bottom hydrodynamic processes. The"first-order" hiatuses evolve removal by erosion of one or several biostratigraphic zones thus disturbing the continuous stratigraphy of a core section. The absence of foraminiferal zone of *Globorotalia crassaformis hessi* (1.47–0.81 Ma) in cores AI-3316 and AI-3318 represents one of the best examples (see Figs. 3, 4 and 7 in Chap. 7). The deep erosion event documented by the high-resolution seismic profiling at the SU-3/SU-4 boundary, reflecting the dramatic change in contourite deposition, also belongs to the first order hiatuses (see Chapters 5, 9 and Sect. 3). We argue that the origin of the"first-order" hiatuses might be related to strong intensification of contour bottom currents resulting from changes in the Lower Circumpolar Deep Water (LCDW) production and delivery to the Ioffe

Drift area from the Antarctic region. In turn, the rough correlation of the major interval with the numerous"third-order" hiatuses between the core sections suggests a correspondence to regional (or even global) paleoceanographic event(s) probably related to changes in the overall velocities of the LCDW flowing out from the Vema Channel and/or its volume over the Ioffe Drift (Fig. 2). Alternatively, the drastic erosion might result from periodic events of enhanced bottom-current velocities according to modeling results by Thran et al. (2018).

Following the earlier ideas (Esin 2003; Esin et al. 2018), we believe that erosion processes accelerate if the dynamically active near-bottom layer is represented by viscous suspension instead of pure sea water. The suspension movement close to the sediment surface driven by gravity force and/or by overflowing currents is able to re-suspend the sediments, thus progressively increasing concentration of the solid phase and eroding ability up to total retardation that means a poorly sorted sediment mass deposition. These processes result in local hiatuses of second- and third-orders, as well as in deposition of poorly sorted contourites.

The bottom-currents activity along with a specific role of near-bottom suspension flows also control layering of the Ioffe Drift contourites, because the layered sedimentary structure is mainly generated by variations in grain size distribution, in turn depending on bottom-current velocities.

Layered sedimentary structures at various scales are typical of contourites (Martín-Chivelet et al. 2008). By contrast, pelagic deposits commonly show a rather homogeneous structure or thick layering, with transitional boundaries between the layers reflecting principal changes in the global scale paleoenvironment. The layering already noted in the visual description of our cores apparently depends on the relationship between sand-size foraminiferal tests, fine silt-size nannofossils and fragments of both. The origin of clay-size (micritic) calcareous crystallites was not ascertained, but their behavior in the mobile bottom water seems to be similar to that of dispersed (non-coagulated) clay particles, which do not settle from suspension even in a slow bottom current. Very thin flaky nannofossils and their fragments can float indefinitely, together with clay particles, to be transported far from their resuspension site, as frequently shown by findings of nannofossil species in filtrated deep-water suspended-matter samples collected from various regions of the Atlantic (Biscaye and Eittreim 1977; Morozov et al. 2010).

Distribution of well-preserved planktic foraminiferal tests (indicating their deposition above the lysocline) together with the strongly dissolved fragments of sortable silt and/or fine-grained sand-size fractions indicating deposition well below the lysocline characterizes a widespread EM2 of the end member modeling (see Figs. 8 and 9 in Chap. 6). This paradox is explained by mixing of the two components derived from different sources during the contourite deposition. Intact foraminiferal tests are moved by bottom currents from close localities at water depths above the lysocline. Fine-grained fragments are thought to be derived from sediments reworking in areas deeper than the lysocline. Nannofossils and *SS* fraction partially with the finest sand is mainly transported to the drift summit from those areas by suspension clouds driven by contour currents and subsequently deposited by selective settling (see Sect. "Lithology of Core Sections").

The study of smear-slides and SEM images reveals that the fine silt is mainly composed of easily floatable calcareous nannofossils. Re-suspended from sediment by bottom-currents or settled out of surface-water phytoplankton to the near-bottom nepheloid layer, they stay in suspension and may be removed from calcareous drifts during their growth. Massive settling of nannoplankton from the surface primary production zone to accumulate as nannofossil ooze is only possible by the pelletal mechanism; that is, rapid settling of sand-size zooplankton pellets composed of nannofossils (Lisitsin 1974, 1978; Kennett 1982). Resuspension of nannofossils after the decay of the pellet coating organic films seems to be a crucial mechanism reducing relative nannofossils content in calcareous contourites compared to that in primary production and pelagic sediments.

Nannofossil productivity and the pelletal mechanism of settling are probably rather weak over the Ioffe Drift area, which is located in a low biological productivity zone within the southern anti-cyclonic gyre (see Fig. 2 in Chap. 1). The restricted supply can hardly explain the domination of foraminiferal material in drift sediments (including the *SS* fraction, coarser than most sediment-forming nannofossils and composed totally of PF fragments). The total re-suspension and washing off nannofossils by bottom-currents seems far more important. The *SS* content in the total <63 μm size fraction and mean sizes of sortable silt-size foraminiferal fragments, increasing the total portion of foraminiferal material, may serve as a measure of bottom-current velocity, as in fine-grained terrigenous contourites (McCave et al. 2017).

Selective settling of calcareous *SS* and coarser, sand-size PF fractions, combined with reworking and partial removal of finer fractions composed mainly of nannofossils, may well represent a major mechanism of calcareous contourites accumulation in the Ioffe Drift. With an accelerating bottom-current velocity, the processes change typically from deposition of poorly sorted fine-grained foram-nanno ooze during Late Pliocene, through nanno-foram ooze with an increasing *SS* content and mean diameter, to well-sorted foraminiferal sand with a minor nannofossil admixture in the late Quaternary. Further currents acceleration results in a change from deposition to total erosion and thus the formation of a hiatus. A rather rare deposition of fine-grained, low-foraminiferal nanno-ooze interlayers in core sections might occur in still bottom-water conditions below eddy centers. Mixing of *SS*-size foraminiferal fragments with nannofossils and clay-size fraction from suspension is generally supported by grain-size data, namely by poor sorting and polymodality of grain-size distributions of individual samples (see Figs. 1, 2, 3, 4, 5 and 6 in Chap. 6).

If so, the grain-size distribution supports the contourite origin of the Ioffe Drift sediments. The layered sedimentary structure shows an alternation of layers of dissimilar grain-size distribution yet without any clear trends (except for abovementioned general coarsening trend) or periodicity (except for bi-gradational cycles in the Upper Pleistocene section of core AI-2436, see Chap. 6). This provides evidence of an irregular pulsating of bottom-current velocities over the core sites on the drift summit. The generally finer-grained Upper Pliocene contourites compared to the most coarse-grained Quaternary ones likely reflect weaker bottom-currents intensity.

Rugged bottom topography (see Figs. 1 in Chaps. 4 and 1 in Chap. 5), irregular relief of major seismic reflectors (see Figs. 3, 4 and 5 in Chap. 5) and lateral variations in the grain-size distribution of assumed simultaneous sedimentary structures (see Figs. 1, 2, 3, 4, 5, 6 and 7 in Chap. 6) collectively suggest unstable bottom-stream configurations rather than changes in the overall strength of the LCDW flow controling the drift's growth. Yet, the poorly studied specific role of suspensions as heavy liquids should also be taken into account in interpretation of the grain-size distribution data.

4.2 Growth of the Ioffe Drift in the Pelagic Realm

The pelagic realm is characterized by a very low terrigenous material supply mainly represented by finest clay size fraction, but likely containing finely dispersed quartz suggesting its remote aeolian origin. In the Ioffe Drift case, this is combined with a low biological productivity in the subtropical anti-cyclonic gyre resulting in a low background pelagic sedimentation rates. The pelagic realm also assumes very low organic matter concentration and accumulation rates which led to oxic environment of the sedimentation and early diagenesis (Kennett 1982; Murdmaa 1987 and references therein).

Transformation of the vertical pelagic settling of planktic foraminifers and calcareous nannofossils packed in pellets, into lateral transport by bottom circulation means generation of calcareous contourites which formed the studied upper cover of the Ioffe Drift above the CCD and possibly the entire drift.

Our hypothesis of the Ioffe Drift's growth is based on two major sources of biogenic calcareous material. The first is a local upward redistribution of intact foraminiferal tests at the bottom surface by ascending suspension streams at the base of contour currents. Mechanisms of such ascending streams are still poorly known, but the main sedimentological result seems to be a partial or total separation of foraminiferal sand from the fine-grained suspension. This leads to a contribution of sand fractions to the drift growth. The second source covers nannofossils and a remote delivery of fine-grained suspended particles, including partially dissolved silt-size foraminiferal fragments in suspension clouds, mainly from the Vema Channel below the lysocline, where erosion (resuspension) takes place due to the high bottom-current velocities of the LCDW. Similar suspension clouds bearing both nannofossils and fine-grained foraminiferal fragments possibly driven by contour currents over the Ioffe Drift summit supply additional fine-grained material from resuspension or erosion in deep channels and depressions (below the foraminiferal lysocline) surrounding the Ioffe Drift (as the core AI-3319, Plate 1). Our multi-proxy data assertained that dissolution by aggressive bottom-waters of Antarctic origin certainly affected the Ioffe Drift lithology below the regional lysocline, which bathymetric position might have varied considerably during the last 3–4 Ma recovered by the cores. New cores from the critical depths are necessary to elucidate the scale and causes of such lysocline depth changes. Measurements of the concentration of suspended particulate material in the water column were carried out at a series of sites, from the Vema Channel to the Florianopolis Fracture Zone Ridge (along the path of the LCDW and AABW flows). They revealed three nepheloid layers occupying different depth intervals (Morozov et al. 2010). The nepheloid layers were traced at water depth ranges of 3,500–4,300 m, 2,200–3,500 and 800–1,500 m, corresponding to the LCDW, NADW and Antarctic Intermediate Water (AAIW), respectively. The volume concentration of suspended matter within the layers varies from 0.1 to 0.5mm^3/l. It is noteworthy that the maximal concentrations of suspended matter were reported from the water mass cores, not their boundaries.

Appearance of several nepheloid layers in the water column represents an anomaly for the pelagic area commonly poor in suspended material. It occurs here due to the intense erosion in the Vema Channel. Moving to the northeast from the Vema Channel, the LCDW flow partially enriched in suspended material (including nannofossils and *SS*-size foraminiferal fragments) meets the Florianopolis Ridge and transforms into a bottom (contour) current able either to deposit *SS* selectively from suspension or to erode older sediments, depending on the contour current velocity (Fig. 4; Frey et al. 2018). A considerable portion of fine silt- and clay-size calcareous particles (micrite) stay in suspension and is removed from the drift system by contour currents, partially responsible for the slow sedimentation rates and relatively increased coarse fraction content as compared to the vertical pelagic sedimentation.

Together with its light yellowish or brownish color and a chemical composition characteristic of oxic early diagenesis, the pelagic realm of the Ioffe Drift's formation is strongly confirmed by the occurrence of hydrogenic ferromanganese nodules. These are especially abundant in the Upper Quaternary deposits of core AI-2436 (Ivanova et al. 2016; see Fig. 1 in Chap. 8). Moreover, rare nodules occur in other cores, including their Upper Pliocene sections, and were abundant in grab samples from the channel at the north-east drift slope below the CCD at stations AI-3653 and AI-3654 during Cruise 52 of the R/V *Akademik Ioffe* (Fig. 3; see Chap. 8).

The behavior of intact PF tests and their large (sand-size) fragments differs dramatically from that of the calcareous silt fraction, due to their much higher settling rate both in clear sea water and in suspension (Berger and Piper 1972). This significantly limits their lateral transport by bottom currents and makes long transport in suspension impossible by bottom currents, resulting in a local supply mainly within the Ioffe Drift itself. Empty intact tests seem to be more readily transportable than fragments of the same size since they are filled with seawater, whereas fragments behave as heavier calcite particles in moving bottom water. The origin of sand-size intact tests might be either pelagic sedimentation within the Ioffe Drift area or the reworking of previously deposited foraminiferal sand. The fragmentation of tests in the sand fraction (>0.1 mm) is mainly related to their porosities (e.g., Sexton et al. 2006) and to dissolution of the more soluble cryptocrystalline calcite cement. Lateral movement of particles took place at the bottom with a rolling mechanism or/and saltation (jumping) over the bottom's surface, depending on bottom-water velocity and turbulence (Allen 1982). Ripple marks of various scales and cross-laminated sedimentary structures are typical of this transport mechanism, confirming the contourite nature of the sediments (Stow and Faugères 2008; Rebesco et al. 2014).

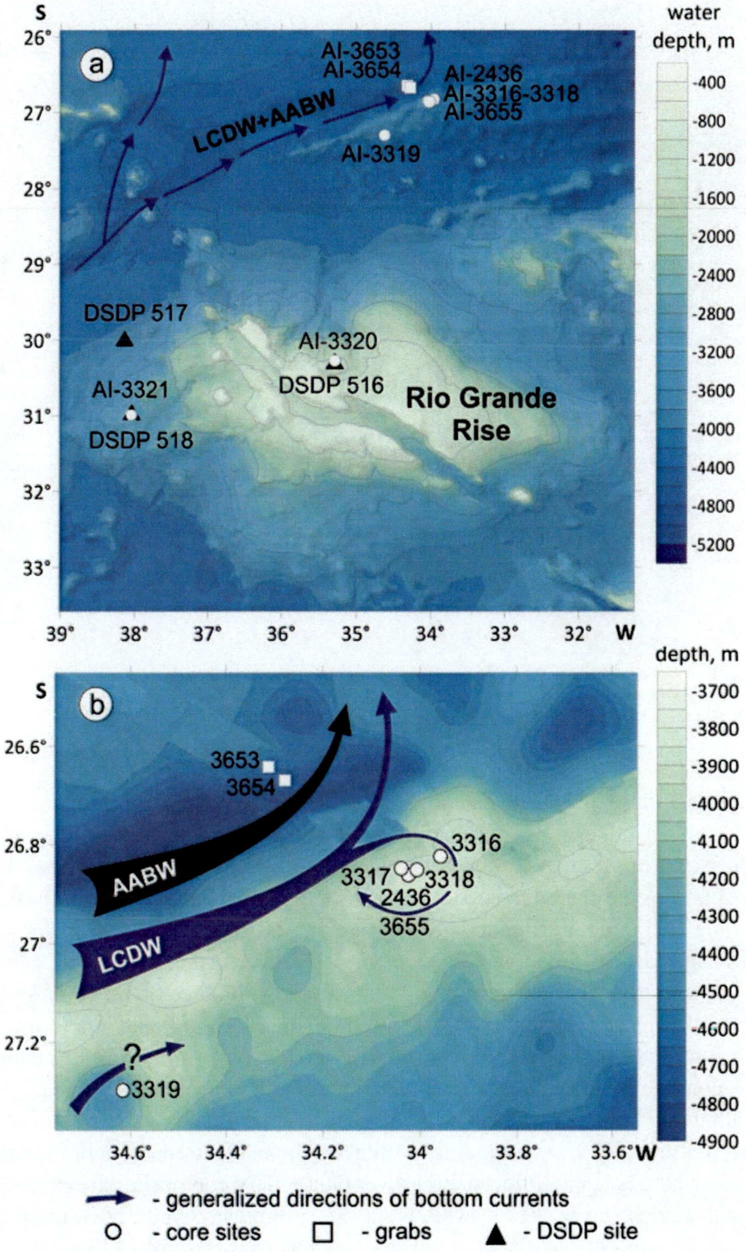

Fig. 3 **a** Modern topography, bottom-water circulation, location of the core/grab sites and DSDP sites in the Ioffe Drift–Rio Grande Rise area; **b** bottom-water circulation with anti-cyclonic gyre in the Ioffe Drift area

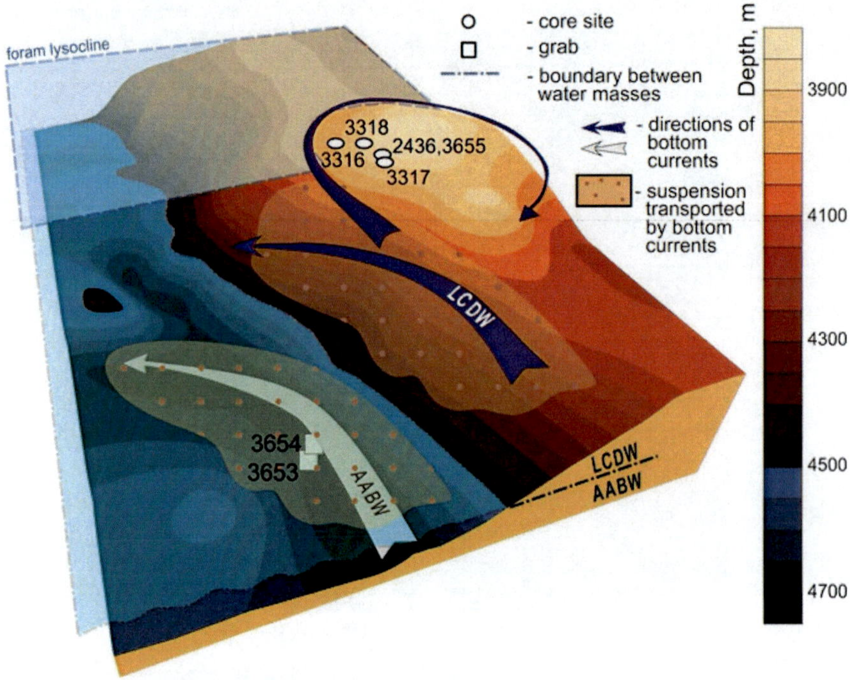

Fig. 4 Sketch of the Ioffe drift formation (modified from Ivanova et al. 2016)

The pelagic realm of the calcareous contourite drift growth presupposes a very low supply of terrigenous material, comprised only of fine-grained clay particles transported from remote continental sources (e.g., Murdmaa 1987). A contribution of aeolian dust to the terrigenous material budget is typical of pelagic sediments, including the minor terrigenous admixture in the high-calcareous pelagic ooze. Further special investigations are necessary to prove its presence in the calcareous oozes of the Ioffe Drift; however, XRF data already obtained on the high Si/Al ratio (up to ten) suggest that finely dispersed aeolian quartz from Patagonia (see Chap. 6) might contribute a considerable portion of the terrigenous admixture of the calcareous ooze, along with the fluvial terrigenous material from the South America.

Rather rare, thin, darker brown clayey interlayers (with calcium carbonate content <80%, minimum 73%) occur mainly within the Upper Quaternary sections of the drift summit, whereas Upper Pliocene to Lower Quaternary contourites are exceptionally high calcareous (up to 99% $CaCO_3$). As the summit lay well above the lysocline during the Late Quaternary (Fig. 4), we suggest that the fall in biogenic $CaCO_3$ content relates to an increase in terrigenous material supply, including that of aeolian dust.

Two bi-gradational contourite cyclites deposited during the last 0.28 Ma and distinguished in the upper 4 m of core AI-2436 (Fig. 1 in Chap. 6) were not found in core AI-3655 (Fig. 6 in Chap. 6) retrieved almost at the same location. Such

Plate 1 SCAN images demonstrating stages of calcareous micro- and nannofossil dissolution: **A–D** Gradual increase in PF dissolution; **E–F** Nannoplankton dissolution

a sharp facies change at a distance not more than about 100 m possibly reflects narrow bottom-current stream which was more erosive at site AI-3655 and continuously accumulative at site AI-2436. Further investigations are necessary to solve this paradox.

Unlike the drift summit and upper slope, deposits with a lower calcium carbonate content were recovered at the base of the Florianopolis Fracture Zone Ridge slope to the west-southwest of the drift by core AI-3319 (water depth 4,066 m; see Fig. 3), owing to carbonate dissolution between the oscillating lysocline and CCD levels. The lysocline depth, in turn, was related to the rate of AABW production and its depth in the area (i.e., Barker et al. 1983b; Hodell et al. 1983). In the lower part of the core, the preservation of PF tests is extremely poor and fine-grained PF fragments dominate over intact tests. Poor PF preservation in the lower part of core AI-3319 is in line with the suggestion by Turnau and Ledbetter (1989) that the upper AABW boundary could have reached DSDP Site 518 (present water depth 3,944 m) by 2.6–2.5 Ma. During the Pliocene–Quaternary variations in the lysocline and calcite compensation depths of the order of several hundred meters have been suggested by several studies, notably in the South Atlantic (e.g., Melguen and Thiede 1974; Shor et al. 1983; Volbers and Henrich 2004).

4.3 Paleoceanographic Context of the History of the Ioffe Drift

The beginning of the Late Pliocene–Quaternary stage of the Ioffe Drift's history roughly coincides with the final closure of the Central American (Panamanian) Seaway (e.g., Keigwin 1982; Schmittner et al. 2004; Jain and Collins 2007; Ivanova et al. 2020) and the strengthening of the deep convection in the North Atlantic by 3.6 Ma (Haug and Tiedemann 1998). Several authors have suggested climatic control over the AABW/LCDW production and distribution during the Neogene–Quaternary (e.g., Barker et al. 1981; Hodell et al. 1983; Kennett and Barker 1990; Berger and Wefer 1996; Hall et al. 2001; McCave and Hall 2006; Ivanova 2009, and references therein). In the northern Antarctic and Subantarctic regions, an increased formation of sea ice favoring the deep convection has been suggested during the Late Pliocene, from 3.2–3 to ~2.5 Ma (Hodell and Venz 1992). This interval was characterized by unstable but relatively warm conditions (Turnau and Ledbetter 1989; Hodell and Venz 1992). The overall strengthening of the Atlantic meridional circulation has recently been ascertained for about the same interval of 3–2.5 Ma (Karas et al. 2017).

The onset of climatic cooling (Shackleton and Opdyke 1977; Prell 1985; Hodell et al. 1983) and the expansion of the West Antarctic ice sheet (Ciesielski and Grinstead 1986) at about 3.2 Ma were followed by a major increase in the northern hemisphere glaciation at 2.7 Ma (Haug and Tiedemann 1998; Lisiecki and Raymo 2005). The introduction of highly saline Mediterranean Outflow Water into the Atlantic at ~2.9 Ma (Loubere and Moss 1986) favored the NADW expansion to the South Atlantic as far as 30°S, with a further deepening to about 3,900 m at ~2.7 Ma (Turnau and Ledbetter 1989).

In the Rio Grande–Vema Channel area, relatively low paleospeeds indicating a sluggish deep circulation are estimated from 3.2 to 2.9 Ma, based on the dominance of relatively fine silt particle sizes, while a reduced paleospeed of AABW and increased paleospeed of deep and intermediate waters are proposed from 2.9 to 2.6 Ma (Turnau and Ledbetter 1989). The short-term events of increased paleospeed at all depths are suggested to have occurred at 3.15–3.10 2.85 and 2.7 Ma. According to our data from the northern slope of the Ioffe Drift, the interval from ~ 4 to 2.59/2.51 Ma was generally characterized by the accumulation prevalence over erosion (Fig. 2). This allows us to assume a rather sluggish or stable bottom-water circulation in the drift area perhaps interrupted by an event of enhanced erosional activity of the LCDW bottom-current at ~3.6–3.5 Ma.

By 2.6–2.5 Ma, a significant volume of southern-origin bottom-water was introduced into the Rio Grande area (Turnau and Ledbetter 1989). This event marks the base of the large erosional interval at 2.59/2.51 Ma in the Ioffe Drift area (Fig. 2 and see Chap. 9). During the latter interval, at about 2.4 Ma, the deep-water stratification became similar to the present system, according to Turnau and Ledbetter (1989).

The Pleistocene epoch is known to be characterized by pronounced glacial–interglacial variability, notably in the pattern and strength of the Atlantic meridional overturning (e.g., Duplessy et al. 1988, 2007; Hall et al. 2001; McManus et al. 2004; Ivanova 2009, and references therein). A greater production of AABW/Circumpolar Deep Water (CDW) during the glacials was suggested by the sortable silt studies in the south-east Pacific (Hall et al. 2001; McCave and Hall 2006). Changes in AABW/LCDW production and input to the south-west Atlantic via the Vema Channel should have significantly affected the sediment deposition and erosion in the large area north of the Vema Channel (see e.g., Ovsepyan and Ivanova 2019, and references therein). However, the link between paleoclimate and bottom-water (notably AABW) paleovelocity was found to be quite complicated (Ledbetter 1979, 1984).

Nevertheless, in our multi-proxy data on the Ioffe Drift sediment cores, vigorous LCDW (and probably AABW) flow during the Late Pliocene–Pleistocene is convincingly indicated by several erosional hiatuses (see Chap. 9). This seems to be consistent with the idea that the drift development rather reflects repeated exceptional events than "an overall increase in background current flow" (Thran et al. 2018). As mentioned above, the hiatuses are most common within the interval 2.59/2.51–1.9 Ma (Fig. 2; see also Fig. 9 in Chap. 9). Besides, according to our new data the interval from 1.47/1.6 to 0.81 Ma contains the"long-term" hiatuses of the first-order. The later interval includes the so-called Mid-Pleistocene Transition (MPT) terminated at 0.95–0.7 Ma (Medina-Elizalde and Lea 2005; de Garidel-Thoron et al. 2005; Kleiven et al. 2011; Hasenfratz et al. 2019). According to our data on the Ioffe Drift, the LCDW flow in the area slowed just after the acceleration events, notably between 1.9 and 1.47 Ma, at the end of MPT, probably at ~0.81 Ma and perhaps later, in the Middle Pleistocene, after 0.27 Ma (Fig. 2).

5 Conclusions

The time limits of the Ioffe Drift's prehistory and early history, as reflected in seismic units SU-1 to SU-3 and the lower part of SU-4, cannot be unambiguously defined. The maximum drift's age of about 95–80 Ma is provided by published paleomagnetic data. The drift was influenced by bottom waters of Southern origin throughout its prehistory. We assume that the intensity of the bottom-currents varied significantly, favoring the dominance of either erosion or accumulation on the drift summit and slopes at different time intervals.

During the Late Pliocene–Recent stage, the drift grew in the oligotrophic pelagic realm under conditions of low terrigenous material supply and low biological productivity, however sufficient for the high-calcareous ooze deposition. This environment is represented in the drift's sediment cover (upper unit SU-4) by calcareous microfossils. Multiple erosional hiatuses identified in the six sediment cores studied suggest that erosion by vigorous bottom-currents was especially intensive during the intervals from 2.51/2.59 to 1.9 Ma and from 1.47/1.6 to 0.81 Ma according to the developed calcareous biostratigraphic framework. Otherwise, high-calcareous sediments accumulated with a low sedimentation rate on the drift summit and slopes above the foraminiferal lysocline. Sediments of a lower $CaCO_3$ content and poorer foraminiferal (and nannofossil) preservation were deposited at greater depths on the slopes above the CCD. Therefore, our new data on the Ioffe Drift contribute to better delimit of the timing of increased AABW/LCDW production during the last 3–4 Ma.

Rather uniform parallel stratified SU-4 suggests a hypothesis that its lower half developed in facial conditions similar to those inferred from the multi-proxy studies of cores which recovered its upper half. Although we have no direct data on the age of the lower half of SU-4, it hardly can be older than the Early Pliocene. Tectonic influence on the drift growth was apparently expressed via depth changes in the Drake Passage and Central American Seaway which directly influenced the intensity of rather independent bottom circulation, including contouritic bottom-water erosion/accumulation activity over the Ioffe Drift. We assume that parallel acoustic reflectors in seismic records might be caused by long-term hiatuses. If so, the dramatic accelerations of bottom-currents shown for the upper half of SU-4 likely interrupted the drift growth during the entire seismic unit formation.

Pronounced down-core variations in grain size distribution depend on bottom-water and/or near-bottom suspension dynamics. Unlike the vertical pelagic sedimentation, the contourite process leads to partial or total removal of fine-grained fractions from the initial biogenic material during its lateral transport and reworking by bottom-currents (and perhaps by near-bottom suspension flows) over the drift. Although dynamics of the near-bottom suspension layer is still poorly studied, its role should be taken into account in any adequate theoretical approaches on the contourite sedimentation.

References

Allen JRL (1982) Sedimentary structures, their character and physical basis, vol 1. Elsevier Science

Allen MB, Armstrong HA (2008) Arabia-Eurasia collision and the forcing of mid-Cenozoic global cooling. Palaeogeogr Palaeoclimatol Palaeoecol 265:52–58. https://doi.org/10.1016/j.palaeo.2008.04.021

Bard E, Arnold M, Duprat J et al (1987) Reconstruction of the last deglaciation: Deconvolved records of 5180 profiles, micropaleontological variations and accelerator mass spectrometric ^{14}C dating. Clim Dyn 1:102–112

Barker PF, Carlson RL, Johnson DA, Party SS (1981) Deep-Sea Drilling Project Leg 72: Southwest Atlantic paleocirculation and Rio Grande Rise tectonics. Geol Soc Am Bull 92:294–309

Barker PF, Johnson DA, Carlson RL et al (1983a) Site 516: Rio Grande Rise. In: Initial reports of the deep-sea drilling project, 72. U.S. Government Printing Office

Barker PF, Johnson DA, Carlson RL et al (1983) Site 518: West Flank, Rio Grande Rise. In: Barker PF, Johnson DA, Carlson RL (eds) Initial reports of the deep-sea drilling project, 72. Government Printing Office, Washington, U.S, pp 357–380

Berger WH, Piper DJW (1972) Planktonic foraminifera: differential settling, dissolution, and redeposition. Limnol Oceanogr 17:275–287. https://doi.org/10.4319/lo.1972.17.2.0275

Berger WH, Wefer G (1996) Expeditions into the past: paleoceanographic studies in the South Atlantic. In: Wefer G, Berger WH, Siedler G, Webb DJ (eds) The South Atlantic: present and past circulation. Springer, Berlin, pp 363–410

Biscaye PE, Eittreim SL (1977) Suspended particulate loads and transports in the nepheloid layer of the abyssal Atlantic Ocean. Mar Geol 23:155–172. https://doi.org/10.1016/0025-3227(77)90087-1

Ciesielski PF, Grinstead GP (1986) Pliocene variations in the position of the Antarctic convergence in the southwest Atlantic. Paleoceanography 1(2):197–232. https://doi.org/10.1029/PA001i002p00197

de Garidel-Thoron T, Rosenthal Y, Bassinot FC, Beaufort L (2005) Stable sea surface temperatures in the western Pacific warm pool over the past 1.75 million years. Nature 433:294–298. https://doi.org/10.1038/nature03189

Duplessy J-C, Roche M, Kageyama M (2007) The deep ocean during the last interglacial period. Science 316:89–91

Duplessy J-C, Shackleton N, Fairbanks R et al (1988) Deepwater source variation during the last climatic cycle and their impact on the global deepwater circulation. Paleoceanography 3:343–360

Esin NV (2003) Viscosity of suspension in gravity flows of sediments. Doklady Earth Sci 393(8):1086–1088 (English Translation)

Esin NV, Murdmaa IO, Esin NI, Evsyukov YD (2018) Dynamics of slow suspension flows on the Black Sea abyssal plain. Quat Intern 465:54–62. www.elsevier.com/locate/quaint

Frey DI, Fomin VV, Tarakanov RY et al (2018) Bottom water flows in the Vema channel and over the Santos Plateau based on the field and numerical experiments. In: Velarde MG, Tarakanov RY, Marchenko AV (eds) The ocean in motion: circulation, waves, polar oceanography. Springer, Cham, pp 475–485

Hall IR, McCave IN, Shackleton NJ et al (2001) Intensified deep Pacific inflow and ventilation during Pleistocene glacial times. Nature 412:809–812. https://doi.org/10.1038/35090552

Hall IR, McCave IN, Zahn R et al (2003) Paleocurrent reconstruction of the deep Pacific inflow during the middle Miocene: Reflections of East Antarctic Ice Sheet growth. Paleoceanography 18(2):1040. https://doi.org/10.1029/2002PA000817

Haq BU, Hardenbol J, Vail P (1987) Chronology of fluctuating sea level since the Triassic. Science 235:1156–1167

Hasenfratz AP, Jaccard SL, Martínez-García A, Sigman DM, Hodell DA, Vance D, Bernasconi SM, Kleiven HF, Haumann FA, Haug GH (2019) The residence time of Southern Ocean surface waters and the 100,000-year ice age cycle. Science 363(6431):1080–1084. https://doi.org/10.1126/science.aat7067

Haug GH, Tiedemann R (1998) Effect of the formation of the Isthmus of Panama on Atlantic Ocean thermohaline circulation. Nature 393:673–676

Hodell DA, Kennett JP (1985) Miocene paleoceanography of the South Atlantic Ocean at 22, 16 and 8 Ma. In: Kennett JP (ed) The Miocene ocean. Geol Soc. Am Mem 163:197–236

Hodell DA, Leonard KA et al (1983) Climatically induced changes in vertical water mass structure of the Vema Channel during the Pliocene: evidence from deep-sea drilling project Holes 516A, 517, and 518. In: Barker PF, Carlson RL, Johnson DA et al (eds) Initial reports of DSDP, Leg 72. U.S. Goverment Printing Office, Washington, pp 907–919

Hodell DA, Venz KA (1992) Toward a high-resolution stable isotopic record of the Southern Ocean during the Pliocene–Pleistocene (4.8–0.8 Ma). In: Kennett JP, Warnke DA (eds) The Antarctic paleoenvironment: a perspective on global change, part 1, vol 56 (Antarctic Research Series). American Geophysical Union, Washington DC, pp 265–310

Ivanova EV, Oskina NS, Blyum NS (1989) Migration of the world ocean climatic zones during the Neogene. Oceanology 29(2):249–255 (English translation)

Ivanova EV (2009) The global thermohaline paleocirculation. Springer, Dordrecht

Ivanova E, Murdmaa I, Borisov D et al (2016) Late Pliocene-Pleistocene stratigraphy and history of formation of the Ioffe calcareous contourite drift, Western South Atlantic. Mar Geol 372:17–30. https://doi.org/10.1016/j.margeo.2015.12.002

Ivanova E, Borisov D, Dmitrenko O, Murdmaa I (2020) Hiatuses in the late Pliocene-Pleistocene stratigraphy of the Ioffe calcareous contourite drift, western South Atlantic. Mar Pet Geol 111:624–637. https://doi.org/10.1016/j.marpetgeo.2019.08.031

Jain S, Collins LS (2007) Trends in Caribbean paleoproductivity related to the Neogene closure of the Central American Seaway. Mar Micropaleontol 63:57–74

Karas C, Nürnberg D, Bahr A et al (2017) Pliocene oceanic seaways and global climate. Nat Sci Rep 7:39842. https://doi.org/10.1038/srep39842

Keigwin LD (1982) Isotope paleoceanography of the Caribbean and east Pacific: role of Panama uplift in late Neogene time. Science 217:350–353

Kennett JP (1982) Marine geology. Prentice-Hall

Kennett JP, Barker PF (1990) Latest Cretaceous to Cenozoic climate and oceanographic developments in the Weddell Sea, Antarctica: an ocean drilling perspective. Proc ODP Sci Results 113:937–960

Kleiven HF, Hall IR, McCave IN et al (2011) Coupled deep-water flow and climate variability in the middle Pleistocene North Atlantic. Geology 39(4):343–346. https://doi.org/10.1130/G31651.1

Krasheninnikov VA, Basov IA (1986) Cenozoic stratigraphy of the Southern Ocean. Nauka, Moscow (in Russian)

Ledbetter MT (1979) Fluctuations of Antarctic bottom water velocity in the Vema Channel during the last 160,000 years. Mar Geol 33:71–89

Ledbetter MT (1984) Bottom-current speed in the Vema Channel recorded by particle-size of sediment fine-fraction. Mar Geol 58:137–149

Lisiecki LE, Raymo ME (2005) A Pliocene–Pleistocene stack of 57 globally distributed benthic $\delta^{18}O$ records. Paleoceanography 20:PA1003. https://doi.org/10.1029/2004PA001071

Lisitsin AP (1974) Sedimentation in oceans. Nauka, Moscow (in Russian)

Lisitsin AP (1978) Processes of oceanic sedimentation. Nauka (in Russian), Moscow

Loubere P, Moss K (1986) Late Pliocene climate change and the onset of northern hemispheric glaciation as recorded in the northeast Atlantic Ocean. Geol Soc Am Bull 97:818–828

Martín-Chivelet J, Fregenal-Martínez MA, Chacón B (2008) Traction structures in contourites. In: Rebesco M, Camerlenghi A (eds) Contourites. Developments in sedimentology 60. Elsevier, Amsterdam pp 157–182

McCave IN, Hall IR (2006) Size sorting in marine muds: Processes, pitfalls, and prospects for paleoflow-speed proxies. Geochem Geophys Geosyst 7:Q10N05. http://dx.doi.org/https://doi.org/10.1029/2006GC001284

McCave IN, Thornalley DJR, Hall IR (2017) Relation of sortable silt grain-size to deep-sea current speeds: calibration of the 'mud current meter.' Deep Sea Res Part I 127:1–12. https://doi.org/10.1016/j.dsr.2017.07.003

McManus JF, Francois R, Gherardi J-M et al (2004) Collapse and rapid resumption of Atlantic meridional circulation linked to deglacial climate changes. Nature 428(6985):834–837

Medina-Elizalde M, Lea DW (2005) The mid-Pleistocene transition in the tropical Pacific. Science 310(5750):1009–1012

Melguen M, Thiede J (1974) Facies distribution and dissolution depths of surface sediment components from the Vema channel and the Rio Grande rise (southwest Atlantic Ocean). Mar Geol 17:341–353. https://doi.org/10.1016/0025-3227(74)90096-6

Morozov EG, Demidov AN, Tarakanov RY, Zenk W (2010) Abyssal channels in the Atlantic Ocean. Springer, Netherlands

Müller RD, Roest WR, Royer J-Y et al (1997) Digital isochrons of the world's ocean floor. J Geophys Res Solid Earth 102:3211–3214. https://doi.org/10.1029/96JB01781

Murdmaa IO (1987) Facies of the ocean. Nauka (in Russian), Moscow

Nisancioglu KH, Raymo ME, Stone PH (2003) Reorganization of Miocene deep water circulation in response to the shoaling of the Central American Seaway. Paleoceanography 18:1. https://doi.org/10.1029/2002PA000767

Ovsepyan EA, Ivanova EV (2019) Glacial-interglacial interplay of southern- and northern-origin deep waters in the São Paulo Plateau-Vema Channel area of the western South Atlantic. Palaeogeogr Palaeoclimatol Palaeoecol 514:349–360. https://doi.org/10.1016/j.palaeo.2018.10.031

Pfuhl HA, McCave IN, Schellenberg SA, Ferretti P (2004) Changes in Southern Ocean circulation in late Oligocene to early Miocene time. In: Exon NF, Kennett JP, Malone MJ (eds) Cenozoic paleoceanography and tectonics in the expanding Tasmanian seaway. American Geophysical Union, Geophysical Monograph

Pfuhl HA, McCave IN (2005) Evidence for late Oligocene establishment of the Antarctic circumpolar current. Earth Planet. Science Letter 235:715–728

Prell WL (1985) The stability of low-latitude sea-surface temperatures: an evaluation of the CLIMAP reconstruction with emphasis on the positive SST anomalies. DOE Report TR025. U.S. Department of Energy, Washington, DC

Rebesco M, Hernández-Molina FJ, Van Rooij D, Wåhlin A (2014) Contourites and associated sediments controlled by deep-water circulation processes: state-of-the-art and future considerations. Mar Geol 352:111–154. https://doi.org/10.1016/j.margeo.2014.03.011

Scher HD, Martin EE (2008) Oligocene deep water export from the North Atlantic and the development of the Antarctic Circumpolar Current examined with neodymium isotopes. Paleoceanography 23:PA1205. https://doi.org/10.1029/2006PA001400

Schmittner A, Sarnthein M, Kinkel H et al (2004) Global impact of the Panamanian seaway closure. Eos 85:526. https://doi.org/10.1029/2004EO490010

Sexton PF, Wilson PA, Pearson PN (2006) Microstructural and geochemical perspectives on planktic foraminiferal preservation: "Glassy" versus "Frosty". Geochem Geophys Geosyst 7

Shackleton NJ, Kennett JP (1975) Paleotemperature history of the Cenozoic and the initiation of Antarctic glaciation: oxygen and carbon isotope analyses in DSDP Sites 277 and 281. In: Initial Reports. DSDP 29. Washington, pp 743–755

Shackleton NJ, Opdyke ND (1977) Oxygen isotope and palaeomagnetic evidence for early northern hemisphere glaciation. Nature 270:216–219. https://doi.org/10.1038/270216a0

Shor AN, Jones G, Rasmussen KA, Burckle LH (1983) Carbonate spikes and displaced components at Deep-Sea Drilling Project Site 515: Pliocene/Pleistocene depositional processes in the Southern Brazil Basin. In: Barker PF, Carlson RL, Johnson DA et al Initial Reports of DSDP, Leg 72. U.S. Goverment Printing Office, Washington, pp 885–893

Stow DAV, Faugères J-C (2008) Contourite facies and the facies model. In: Rebesco M, Camerlenghi A (eds) Contourites. Developments in sedimentology, chap 13 60. Elsevier, Amsterdam, pp 223–256

Stow DAV, Smillie Z, Pan J, Esentia I (2019) Deep-sea contourites: sediments and cycles. In: Encyclopedia of ocean sciences. Elsevier, pp 111–120

Thomas DJ, Via RK (2007) Neogene evolution of Atlantic thermohaline circulation: perspective from Walvis Ridge, southeastern Atlantic Ocean. Paleoceanography 22. https://doi.org/10.1029/2006PA001297

Thran AC, Dutkiewicza A, Spence P, Müller RD (2018) Controls on the global distribution of contourite drifts: insights from an eddy-resolving ocean model. Earth Planet Sci Lett 489:228–240. https://doi.org/10.1016/j.epsl.2018.02.044

Trauth MH, Sarnthein M, Arnold M (1997) Bioturbational mixing depth and carbon flux at the seafloor. Paleoceanography 12(3):517–526

Turnau R, Ledbetter MT (1989) Deep circulation changes in the South Atlantic Ocean: response to initiation of Northern Hemisphere glaciation. Paleoceanography 4:565–583

Via RK, Thomas DJ (2006) Evolution of Atlantic thermohaline circulation: early Oligocene onset of deep-water production in the North Atlantic. Geology 34:441–444

Volbers ANA, Henrich R (2004) Calcium carbonate corrosiveness in the South Atlantic during the last glacial maximum as inferred from changes in the preservation of Globigerina bulloides: a proxy to determine deep-water circulation patterns? Mar Geol 204:43–57. https://doi.org/10.1016/S0025-3227(03)00372-4

Wetzel A, Werner F, Stow DAV (2008) Bioturbation and biogenic sedimentary structures in contourites. In: Rebesco M, Camerlenghi A (eds) Contourites. Developments in sedimentology, chap. 11 60. Elsevier, Amsterdam, pp 183–202

Conclusions and Perspectives

Elena Ivanova, Ivar Murdmaa and Dmitrii Borisov

The main task of this monograph has been to demonstrate that the calcareous contourite drift in the southern Brazil Basin, discovered in 2010 and named the Ioffe Drift, has been formed by alternating phases of erosion and accumulation due to the changing intensity of the AABW/ LCDW outflow from the Vema Channel. The following conclusions can be inferred from the present study.

According to results of seismic profiling and bathymetric surveys, the Ioffe Drift comprises an asymmetric, lens-like, mounded depositional body overlying the central part of a large ridge extending in the WSW–ENE direction within the Florianopolis Fracture Zone. The thickness of the deposits overlying the ridge exceeds 200 m. The seismic structure of the drift was studied in detail only for the upper 80 m. Four seismic units were determined within this interval, separated by angular unconformities. The overall geometry, orientation, erosional features (moats at the base of the mounded basement outcrops) expressed in the body morphology, together with the shape of the seismic units and internal unconformities, indicates the contourite origin of the Ioffe Drift. The defined seismic units demonstrate a character and configuration of reflectors corresponding to phases of alternating contourite deposition and erosion by the weakening and intensification of bottom currents, probably followed by prevalent conditions of pelagic settling. Evidence of drastic erosion is traced over the entire drift structure. The major unconformities separating the units correspond to the most prominent erosional events in the drift's history. An estimation of its age is possible only for the uppermost unconformity, probably related to the Late Miocene–Pliocene boundary.

The multi-proxy study of six cores from the Ioffe Drift area, notably grain-size distribution, reveals the contourite origin of upper sediment cover. The lithology of the calcareous Ioffe Drift differs in composition (totally low-magnesium calcite) from its terrigenous analogs of similar grain-size distribution and the initial biomorphic shape (planktic foraminiferal tests, calcareous nannofossils) of the sediment-forming particles. The behavior of both the intact biomorphic particles and their fragments

I. Murdmaa and E. Ivanova (eds.), *The Ioffe Drift*, Springer Geology,
https://doi.org/10.1007/978-3-030-82871-4

resembles that of terrigenous material although differs considerably in shape of biomorphic particles of equal size. The revealed lines of evidence ascertain that the sedimentological regularities of contourite theory are equally valid for calcareous drifts.

Common features that are apparent in the grain-size distribution were documented. The silt fractions dominate, with a considerable contribution of the sortable silt (*SS*). The relationship between *SS* concentration in the total <63 μm size fraction and *SS* mean sizes suggests an influence by bottom-current velocities on the contourite deposition. The commonly poor sorting characteristic of contourites in our case is related to bimodal or polymodal individual grain-size distributions which, in turn, are created by differing transport and deposition mechanisms of the main sediment components, namely foraminiferal sand, fine foraminiferal test fragments, mostly of *SS* size, and nannofossils together with clay-sized particles.

The commonly good to perfect foraminiferal preservation in a major part of the core sections studied suggests that the lysocline was generally deeper than 3.8 km during the past ~3–3.5 Ma. However, the foraminiferal lysocline and the calcite compensation depth (CCD) might have been shallower in the Late Pliocene, possibly ~3.5–4 Ma, than today. This follows from the significant to strong foraminiferal dissolution documented in the lower parts of cores AI- 3317 and AI-3319, collected from the present depths of 3.83 km and 4.07 km, respectively.

The biostratigraphic framework of the six sediment cores from the Ioffe Drift area covers the Late Pliocene–Late Pleistocene (to Recent). Core AI-3317 from the northern slope of the drift spans the last ~4 Ma; that is, it penetrated older sediments than the other cores. The biostratigraphic study allowed the identification of Late Pliocene–Quaternary foraminiferal (in all six cores studied) and nannofossil (in four cores studied) zones and their correlation to those previously reported by several authors on the neighboring Rio Grande Rise (Berggren et al. 1983a, b, 1995; Barash et al. 1983; Dmitrenko 1987; Barash 1988).

The reduced thickness of biostratigraphic zones in the Ioffe Drift area relative to those at DSDP Site 516 on the Rio Grande Rise, along with the absence of five nannofossil and four foraminiferal zones from some sections, indicates several long-term hiatuses, notably in the Upper Pliocene–Middle Pleistocene part of the section. Two major stratigraphic gaps of up to ~0.6–1 Ma-long suggest the occurrence of first-order erosional hiatuses.

The integration of the biostratigraphic data and the volume magnetic susceptibility (MS), color reflectance L* and X-ray Fluorescence (XRF) data has enabled us to identify erosional hiatuses confidently, notably in the interval from 2.51/2.59 to 1.9 Ma, associated with the reorganization of deep-water circulation since the Pliocene/Pleistocene boundary according to Turnau and Ledbetter (1989). Some (probably short-lived) younger and older hiatuses have been also tentatively identified from five cores in which multi-proxy analyses were performed. Among them, the second-order hiatuses could be identified by rather transitional faunal/floral changes at the stratigraphic boundaries, along with the abrupt high-amplitude changes in volume MS, color reflectance L* and XRF data. The third-order hiatuses are generally well represented in five records of the parameters mentioned above; however, they cannot be identified by biostratigraphic zonation.

The multiple hiatuses inferred in all cores studied from the Ioffe Drift area have resulted in much slower growth of the Ioffe Drift than drifts in the North Atlantic, for instance, and have hindered any reliable estimate of sedimentation rates. On the neighboring Rio Grande Rise, published DSDP data and estimates for our dated core AI-3321 from the rise slope suggest mean sedimentation rates of about 0.6–1 cm/ka. Even if erosion is not taken into consideration, these values are higher than those in the Ioffe Drift area. The overall relatively slow growth of the drift is mostly explained by extensive erosion by bottom currents.

The intensity of the AABW/LCDW flow through the Vema Channel, notably its eastern branch, exhibited strong variations during the Late Pliocene–Quaternary, significantly affecting sedimentation and resulting in alternating calcareous contourite accumulation and erosion events in the Ioffe Drift. In general, bottom currents were more vigorous during the aforementioned intervals in Late Pliocene–Early Pleistocene, from 2.51/2.59 to 1.9 Ma and especially in the Early-Middle Pleistocene, from 1.6 to 0.81 Ma, probably reflecting increased AABW/LCDW production in the Southern Ocean. The LCDW flow slowed probably from ~1.9 to 1.6 Ma, after the Mid-Pleistocene Transition (i.e., after ~0.8 Ma, in our records) and then possibly after about 0.27 Ma, maintaining contourite deposition under rather stable oceanographic conditions.

In future, our knowledge of the Ioffe Drift and our understanding of its controlling mechanisms can be significantly improved by detailed investigation of the bottom relief (using multibeam profilers and side-scan sonars), as well as the deep seismic structure of the drift (using continuous seismic profiling). Detailed high-resolution seismoacoustic survey has to be expanded to investigate the Upper Miocene–Pliocene acoustic structure of the drift's upper sediment cover and to prepare sites for coring.

It is necessary to retrieve at least 10–15 new cores to understand the processes of the drift growth in terms of sedimentology and biostratigraphy. Wedging out structures of the uppermost seismic unit SU-4, which upper half has been recovered on the drift summit by studied cores, represents a special interest. Sedimentary facies mapping by coring and supplementary grab sampling will provide a valuable new tool for investigation of erosion/deposition processes depending on the local bathymetry.

The study of near bottom suspended material, its concentration and composition is an important issue of future work on the Ioffe Drift and its surroundings. Possible variations in the non-calcareous components supply into the drift sediments composition revealed by mineralogical analyses help to solve regional to global problems of eolian and fluvial sedimentary material sources and transport ways.

The water column structure should be studied by CTD measurements (of conductivity, temperature and pressure of seawater) to clarify the location of the water mass boundaries. In particular, moored current meters, LADCP (Lowered Acoustic Doppler Current Profiler) measurements and numerical modeling of current velocities may shed light on the circulation patterns over the drift and adjacent areas. The study of vertical and horizontal fluxes of suspended matter in the water column is expected to clarify the mechanisms of sediment transport and deposition. The analysis of the cores using computer tomography will confirm the inferred hiatuses (or perhaps find new ones) and investigate ichnofossils, an important source of information on changes in bottom-water dynamics. An inter-correlation of several independent proxy records applied in this study serves as the basis of future elaboration of theoretical approaches explaining the AABW/LCDW circulation patterns resulted in the revealed extensive hiatus formation characteristic of the Ioffe Drift.

As the basaltic basement is probably exposed at the base of the steep northern slope of the drift and rock fragments may occur in the northern fault channel bordering the drift, dredging may yield basalt fragments for use in its dating. However, carrying out a systematic coring program directly related to seismoacoustic survey leaves the major task for future expeditions.

In the global context, the modern contourite paradigm assumes that it is possible to apply knowledge about contourite drifts that have been discovered recently on the seafloor to facies models for comparative studies of potential oil- and gas-bearing fossil analogs (e.g.,Viana et al. 2007; Rebesco et al. 2014). The calcareous Ioffe Drift, formed in a typical oligotrophic pelagic realm characterized by a low marine organic matter content and very slow average sedimentation rates, due to numerous hiatuses, hardly serves as a model of a potential hydrocarbons collector; however, its investigation may be valuable in terms of wider issues of the calcareous (foraminiferal-dominated) contourite accumulation that may well occur in geological settings more conducive to the generation of hydrocarbon deposits.

References

Barash MS (1988) Quaternary paleooceanology of the Atlantic Ocean. Nauka (in Russian), Moscow

Barash MS, Oskina NS, Blyum NS (1983) Quaternary biostratigraphy and surface paleotemperatures based on planktonic foraminifers. In: Barker PF, Johnson DA, Carlson RL (eds) Initial reports of the deep-sea drilling project, 72. Government Printing Office, Washington, U.S, pp 849–869

Berggren WA, Aubry MP, Hamilton N (1983a) Neogene magnetobiostratigraphy of deep-sea drilling project, site 516: Rio Grande Rise, South Atlantic. In: Barker PF, Carlson RL, Johnson DA (eds) Initial reports of deep-sea drilling project 72. Government Printing Office, Washington, U.S, pp 675–713

Berggren WA, Hamilton N, Johnson DA, Pujol C, Weiss W, Cepek P, Gombos AM Jr (1983b) Magnetobiostratigraphy of deep-sea drilling project 72, Sites 515–518: Rio Grande Rise (South Atlantic). In: Barker PF, Carlson RL, Johnson DA (eds) Initial reports of deep-sea drilling project 72. Government Printing Office, Washington, U.S, pp 939–948

Berggren WA, Hilgen FJ, Langereis CG et al (1995) Late Neogene chronology: new perspectives in high-resolution stratigraphy. Geol Soc Am Bull 107:1272–1287. https://doi.org/10.1130/0016-7606(1995)107%3c1272:LNCNPI%3e2.3.CO;2

Dmitrenko OB (1987) A detailed zonal scale of the quaternary bottom deposits based on coccoliths (on the Rio Grande Rise in the Atlantic Ocean). Oceanology 27(3):460–464 (in Russian with English trans)

Rebesco M, Hernández-Molina FJ, Van Rooij D, Wåhlin A (2014) Contourites and associated sediments controlled by deep-water circulation processes: state-of-the-art and future considerations. Mar Geol 352:111–154. https://doi.org/10.1016/j.margeo.2014.03.011

Schmittner A, Sarnthein M, Kinkel H et al (2004) Global impact of the Panamanian seaway closure. Eos 85:526. https://doi.org/10.1029/2004EO490010

Turnau R, Ledbetter MT (1989) Deep circulation changes in the South Atlantic Ocean: response to initiation of Northern Hemisphere glaciation. Paleoceanography 4:565–583

Viana AR , Almeida W, Nunes MCV, Bulhões EM (2007) The economic importance of contourites. Geol Soc London, Spec Publ 276:1–23. https://doi.org/10.1144/GSL.SP.2007.276.01.01

Index

I. Murdmaa and E. Ivanova (eds.), *The Ioffe Drift*, Springer Geology,
https://doi.org/10.1007/978-3-030-82871-4

C

D

E

F